Nicoletta Sala
Editor

Complexity and Fractality in Human Change and Beyond

nova
science publishers
www.novapublishers.com

DOI: https://doi.org/10.52305/CFIT2710

Additional color graphics may be available in the e-book version of this book.

Library of Congress Cataloging-in-Publication Data

ISBN: 979-8-89530-575-1 (Hardcover)
ISBN: 979-8-89530-620-8 (e-Book)

Published by Nova Science Publishers, Inc. † New York

Contents

Preface

"The essence of tyranny is the denial of complexity."
- Jacob Burchardt (Swiss Historian, 1818-1897).

Figure 1. AI-generated fractal image based on suggestions by Nicoletta Sala.

This is the first book in the new series, Human Dynamics: Multidisciplinary Complexity Science, edited by Franco F. Orsucci and Nicoletta Sala.

This series of topical volumes offers an authoritative resource for understanding and applying the concepts, tools, and measures of complexity theory to analyze complex systems across all fields of science, with a particular focus on human dynamics. Many phenomena across all scales in science exhibit the characteristics of complex systems. They can be fully understood only through transdisciplinary perspectives, theories, and tools that include self-organization, dynamical systems, turbulence, catastrophes, instabilities, nonlinearity, stochastic processes, chaos, neural networks, cellular automata, adaptive systems, genetic algorithms, quantum theories, and more. All these seemingly diverse phenomena and structure formations have essential features and underlying structures in common. These profound structural similarities can be exploited to transfer analytical methods and understanding from one field to another. This unique work will extend the influence of complexity and system science to a broader audience. Ecosystems, the human psyche-soma, ant colonies, the internet, and economic networks are all complex systems displaying collective behavior and emergence beyond the sum of their parts. Complexity science systematically investigates these emergent phenomena and stretches across and beyond disciplines.

According to the editorial series, this volume collects works from economics, sports psychology, architecture, number theory, and urban planning.

During the preparation of this book, Gabriele A. Losa passed away. He gained international recognition through pioneering works on fractal geometry applied to cell biology and human cancers. He was a Member of the European Academy of Sciences. BSc in Natural Sciences and PhD (1972) from the Faculty of Biology and Medicine of Lausanne University. Postdoctoral Fellow at the Department of Cell Biology and Clinical Cancer Research Institute of Berne University, the Institute of Cellular Pathology of Louvain University, and the Memorial Sloan Kettering Cancer Center of New York. Founder of the Laboratory of Cellular Pathology and the Institute for Scientific Interdisciplinary Studies, Locarno (Switzerland). He taught cell biology at the Lausanne, Turin, Bari, and Naples universities. His academic collaboration with the Department of Mathematical Physics at the University of Ulm focused on investigating the role of fractal geometry in biology and medicine and developing mathematical software for fractal analysis. Founder, chairman, and editor of the International Symposium "Fractals in Biology and Medicine," which began in 1993 in Locarno-Ascona, Switzerland. He had the great privilege of being a close friend of Professor Benoît Mandelbrot, to whose memory the entire series of proceedings volumes, Fractals in Biology and Medicine, is dedicated.

His journey through life and research formed intricate, fractal patterns that helped shape new fields in the biological sciences.

To commemorate Gabriele Losa, the first chapter of this book will host the English translation of one of his works entitled "Comportamenti Frattali in Biologia Umana e Ambientale," previously published in Italian on *Temi*

Interdisciplinari di Natura Scientifica: Dinamiche Complesse e Strutture Frattali, anno 4, vol. 4, pp. 37-51 (thanks to the kind permission of Professor Danilo Merlini, PhD, Director of CERFIM Locarno).

Locarno, 15 December 2024

Nicoletta Sala, PhD
Institute for Scientific and Interdisciplinary Studies,
ISSI, Locarno, Switzerland
Research Institute in Arithmetic Physics, IRFIA, Locarno, Switzerland
Institute for Complexity Studies, Rome, Italy
Editor

Chapter 1

Fractal Behaviors in Human and Environmental Biology*

Gabriele Losa†
Laboratorio di Patologia Cellulare, Istituto Cantonale di Patologia, Locarno, Switzerland
University of Lausanne, Lausanne, Switzerland

Abstract

This work aims to introduce some interesting connections between fractal geometry in human and environmental biology.

Keywords: biology, complexity, fractal, fractal dimension, human and environmental biology

1. Introduction

Most tissues, cells, and their intracellular organelles show irregular shapes and discontinuous and complex morphogenetic processes connected with the functional role. In reality, irregular objects' morphological and infrastructural diversity is casually ignored and rarely measured since they are assimilated

---

*This work is a translation of Losa, G. (1994), "Comportamenti Frattali in Biologia Umana e Ambientale," which appeared on *Dinamiche Complesse e Strutture Frattali in Biologia Umana Ambientale*, Edizioni ISSI, 1994, 4(4), 37-51. With the kind permission of Prof. Dr. Danilo Merlini (President of CERFIM and ISSI, Locarno Switzerland). Translation by Prof. Dr. Nicoletta Sala (ISSI, Locarno, Switzerland).

† Corresponding Author's Email: nicolettasalavb@gmail.com

In: Complexity and Fractality in Human Change and Beyond
Editor: Nicoletta Sala
ISBN: 979-8-89530-575-1

into regular and ideal Euclidean geometric bodies, such as spheres, ellipses, triangles, etc.

Arbitrary simplification of any real form entails the inevitable loss of dimensional geometric and morphological spatial information and unreliable data acquisition when proceeding with morphometric measurements of two-dimensional representations of cells and subcellular structures. In this regard, it is commonly used to refer to the profound divergence between the data regarding the surface and volume density of hepatocyte membranes obtained in various studies (Loud, 1968; Weibel, 1969; Reith, 1973). This discrepancy should not be traced back to different factors such as the difference in age, nutritional status, variability of cell strain, or even to the procedures of fixation, tissue processing, and quantitative evaluation but rather to the use of scales of measurements of different magnification levels.

Indeed, recent studies have highlighted that the magnification used for morphometric measurements significantly determines the values of dimensional parameters and subcellular ultrastructures (Paumgartner et al., 1981; MacAulay & Palcic, 1990; Nonnenmacher, 1994).

Therefore, the length, surface area, and other dimensional quantities of irregular objects must be evaluated with adequate morphological and morphometric procedures and scales and not chosen a priori, as illustrated below. As regards the use of systems dedicated to automated image analysis at the submicroscopic level, it should be noted that it will be limited, at least until it is possible to make use of a reliable procedure for recognizing the tone of gray that ultrastructures take on in electron microscopy (Rigaut, 1991).

2. Characteristics and Properties of Fractal Objects

The term fractal derives from "fractus," which means interrupted irregular and fractal geometry serves to mathematically describe those irregular, structurally complex natural objects and phenomena characterized by a fractal dimension of non-integer value, different from the ordinary, topological dimension, which represents the objects regular and ideal (Euclidean). It follows that the scope of the topological dimension is "resized" and placed in a context in which the fractal dimension takes on a universal generalizing connotation, according to the theory of fractal geometry coined by Benoit Mandelbrot (Mandelbrot, 1983). In addition to irregularity, the fractal object is characterized in a physical sense by the property of self-similarity or internal homothety.

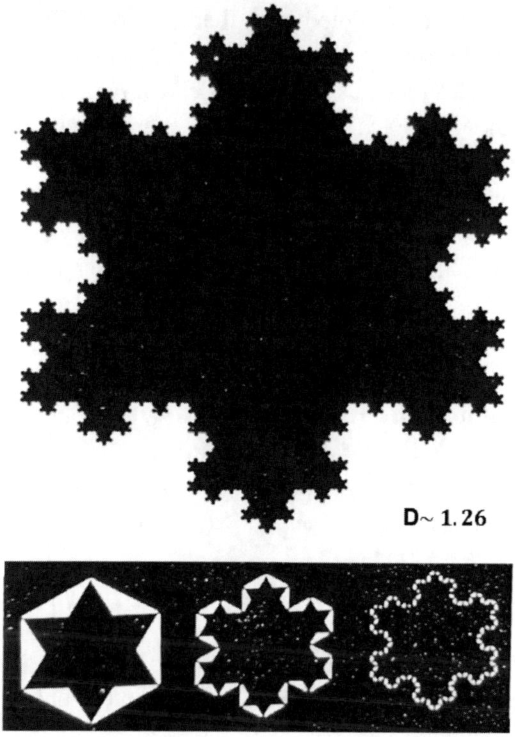

Figure 1. Von Koch snowflakes.

In other words, an object is fractal when its shape or fragment reflects its whole or offers similar structural properties for any scale of measurement adopted. This principle of geometric and/or statistical self-similarity, whereby the degree of complexity or even the quantity of detail remains constant at any scale of observation or magnification (invariance of shape). This is fundamental in defining a fractal object. The intimate meaning is not of immediate meaning, especially since to establish the fractal dimension of an object (for example, the length of a cellular perimeter, the surface of a subcellular organ, etc.), one proceeds with successive measurements using rulers of variable scale. The measurement will be different each time, and the smaller the ruler, the more significant the result will be, while the degree of complexity or irregularity will remain unchanged. It can be deduced from a mathematical point of view that if the ruler is progressively reduced according to a specific ratio, the length of the perimeter of an irregular object grows inexorably, and the ruler tending to zero, it tends to infinity. Physically, this makes little sense since the perimeter of a real object is always finite.

However, it should be noted that there are fractal, mathematical, or theoretical objects with exact and geometric self-similarity properties for an infinite range of scales (such as the Von Koch curve or snowflake, Figure 1, while natural and biological objects (for example, the contour of the pericellular membrane), being generally fractal constructions generated not by an unlimited process of simple iterations, but rather by an irregular one, shows a statistical and therefore effective self-similarity only for an interval of measurement scales that must be established empirically (Baumann et al., 1994).

3. Determination of the Fractal Dimension

We exemplify this by choosing a line segment AB (Figure 2) whose length L is expressed by the formula

$$L = N \cdot r.$$

where N represents the number of identical parts contained with the help of a ruler of length r, between N, the ruler r, and the dimension D exists a relation that satisfies the equation:

$$N(r) = \left(\frac{L}{r}\right)^{D} \text{ or } D = \frac{\log N(r)}{\log\left(\frac{L}{r}\right)} \quad (1)$$

The fractal dimension D = 1 characterizes an object of topological dimension 1, represented by the straight line; D = 2 characterizes a bidimensional object (surface) dimension, and so on. A non-integer value of D connotes the fractal dimension of an irregular object.

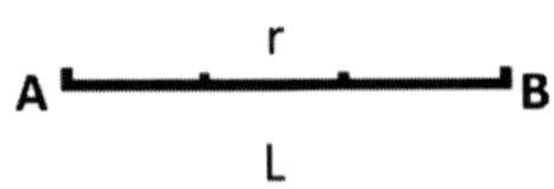

Figure 2. A line segment AB.

What is the difference between a fractal object and an Euclidean object? Let us illustrate it by constructing the von Koch curve (exact or mathematical fractal). To connect the segment AB instead of three, let us divide it into four identical segments of length r, replacing the median segment with two elements constituting an equilateral triangle (Figure 3).

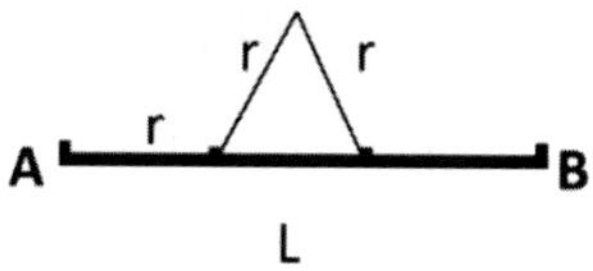

Figure 3. Construction of a von Koch curve.

The length L of the broken line AB is given by the following relationship:

$L(r) = N \cdot r$ where $N = 4$ and $r = \frac{L}{3}$.

$$L(r) = \frac{(4) \cdot L}{3}$$

Replacing in Eq. (1):

$$D = \frac{\log N}{\log\left(\frac{L}{r}\right)} = \frac{\log 4}{\log\left(\frac{L}{\frac{L}{3}}\right)} = \frac{\log 4}{\log 3} = 1.26$$

The use of a scaling ruler, increasingly reduced by the factor of one-third, i.e., by scaling the ratio r, r/L for n iterations, the factor becomes $(1/3)^n$, so by substituting it in Eq. (1) gives the relationship:

$$L(r) = \left(\frac{4}{3}\right)^n \cdot L$$

where for $n \to \infty$, $L(r) \to \infty$.

The nonlinear behavior of a curve or fractal object is described by a relation as follows:

$$L(r) = N \cdot r$$

Replacing N with $(L/r)^D$, we obtain:

$$L(r) = \left(\frac{L}{r}\right)^D \cdot r = \frac{L^D}{r^D} \cdot r = L \cdot \frac{L^{D-1}}{r^{D-1}}$$

$$L(r) = L \cdot \left(\frac{L}{r}\right)^{D-1} \quad (2)$$

where for $r \to 0$ and $D > 1$, $L(r) \to \infty$.

The Eq. (2) translates a power law which is expressed by appropriately transforming and replacing with:

$$\frac{L(r)}{L} = P \text{ (perimeter) and } \frac{r}{L} = r_s \text{ (scaling ruler).}$$

Obtaining:

$$\frac{L(r)}{L} = \left(\frac{L}{r}\right)^{D-1} = \left[\left(\frac{r}{L}\right)^{-1}\right]^{D-1} = \left(\frac{r}{L}\right)^{1-D}$$

Thus

$$P = (r_s)^{1-D} \tag{3}$$

Now, transforming in log-log the Eq.(3) we obtain:

$$\text{Log } P = (1\text{-}D)\log r_s \tag{4}$$

Eq. (4) represents the equation of a straight line with a slope of (1-D), Figure 4, from which the fractal dimension D is deduced.

Therefore, the previous expression is transformed into a log-log and represents the equation of a straight line with slope 1 minus d, from which the fractal dimension D is deduced.

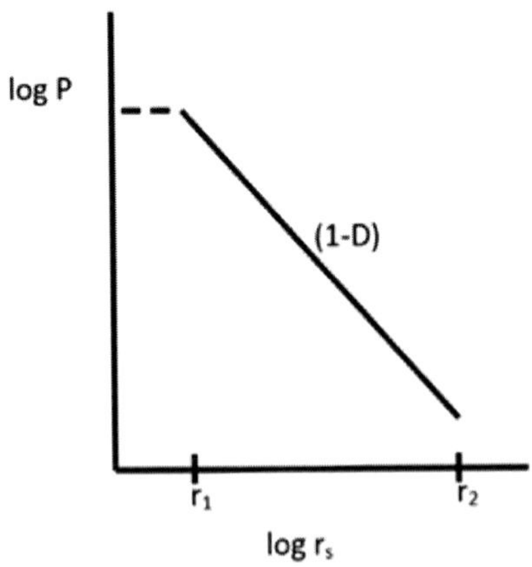

Figure 4. An example of a log-log plot,

In the specific case in which P constitutes the length of the cellular perimeter or a cytoplasmic organelle and r_s the scaling ruler, whose measurement decreases with increasing magnification M used for the micrographs, we obtain $r_s = a \cdot M^{-1}$, where a resolution or effective scaling unit at which the object is observed. Thus, M = 1, $r_s = a$; for $M = 10^3$, $r_s = a \cdot 10^{-3}$ and so on.

Then, Eq. (4) becomes in terms of magnification:

$$\log P = (1\text{-}D) \cdot \log M^{-1} \quad (5)$$

where the length of the profile or perimeter P is a function of the resolution, i.e., the magnification M, which must be included in a scale interval that must be established empirically and within whose physical meaning and self-similarity of the object under examination are contemplated (9). Outside the scale interval, the biological object ceases to be fractal, and the dimension regains its entire topological value (r_1/r_2: interval of scale).

4. Biological Fractals and Statistical Applications

In biology, there are numerous examples of spatial geometric fractals, among which we will mention: a) the extensive external and internal surfaces confined in limited spaces such as cell membranes, alveolar gas exchange membranes, or the leaf limbs of plants; (b) branching systems of vessels or pathways for the rapid conduction and distribution of blood, oxygen, fluids, metabolites, etc., (c) morphological and structural cytosystems with their changes during cell proliferation, tumor transformation, and other pathologies related to stress and environmental alterations. The fractal dimension, as well as being suitable for quantitatively describing the value of irregularity, the length of the profile or entity of the surface of biological elements, can also serve as a diagnostic, pathological, and taxonomic discriminant.

The internal membrane of cellular mitochondria belongs to the first group of fractal objects, in which the enzymatic proteins are responsible for the formation of H_2O for oxidation-reduction and the production of energy accumulated in triphosphate nucleotides.

The membrane's length and surface density increase with increasing magnification or experimental resolution. This effect is attributable to the fact that the evaluation of the irregularity of the mitochondrial membrane is linked to the strengthening of the capacity for identification (resolution). In reality, the degree of irregularity of the membrane does not change when the measurement scale changes since it is an irregular and self-similar fractal object in a statistical sense.

The application of fractal analysis to determine the length of the leaf profile served to describe the varieties of vines and other botanical species and to define the functional and phytopathological state (Kaye, 1989).

As part of research carried out to characterize the pathogenetic mechanisms and the biochemical and ultrastructural properties of neoplasms (leukemia and lymph nodes) of the hematopoietic system, the degree of complexity and irregularity of the cellular surface of leukemic elements obtained "*ex vivo*," and of cell lines in lymphoblastoid was analyzed.

The fractal dimension, evaluated with probabilistic methods after digital acquisition of ultrastructural data detectable from two-dimensional planar projections of elements obtained under the electron microscope, provides a quantitative index to define the proliferation stage and the degree of differentiation of leukemic cells. Furthermore, it allows the quantitative description of the shape of a lesion and can, therefore, take on objective diagnostic value. In general, lymphoid elements marked by a very irregular and complex surface profile (hairy leukemia) are distinguished by their fractal dimension (D = 1.32), Figure 5, higher than that expressed by mature lymphocytes T, Figure 6, and immunocompetent (D = 1.20). The fractal dimension (D = 1.11) of leukemic lymphoblasts and lymphoblasts T obtained by stimulation with a mitogen molecule appears to be significantly smaller (Figure 7). These elements are characterized by a high proliferation index but by a smooth and slightly irregular surface profile when compared to previous cytotypes (Losa, 1992; Losa, 1994).

An interesting application of fractal geometry is in the field of industrial environmental hygiene. In coal dust, the burnt fine particles and soot inhaled (Kaye, 1989) are deposited along the respiratory tract, causing lesions and various pathologies of the lung tissue.

It was possible to ascertain that the kinetics and dynamics of particle deposition follow a non-linear and, therefore, unpredictable trend due to the morphological irregularity and heterogeneity of the particles on the one hand and the intervention of two opposing forces on the other, such as the mass of the particle that drags it downwards and the flow of air behind the particle that slows its progress.

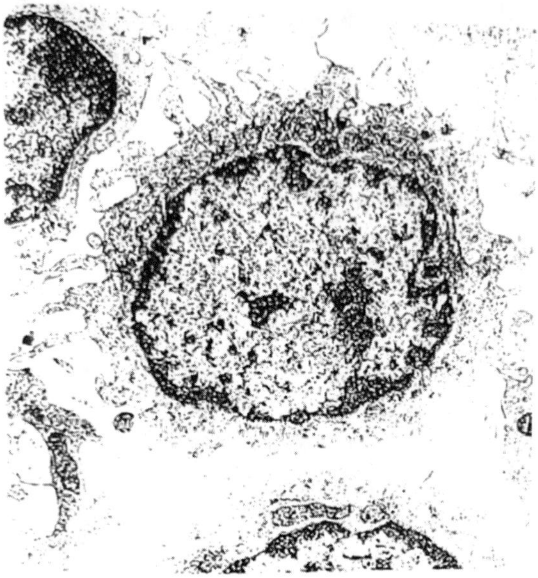

Figure 5. Electron microscope image of a hairy leukemia element (D = 1.32).

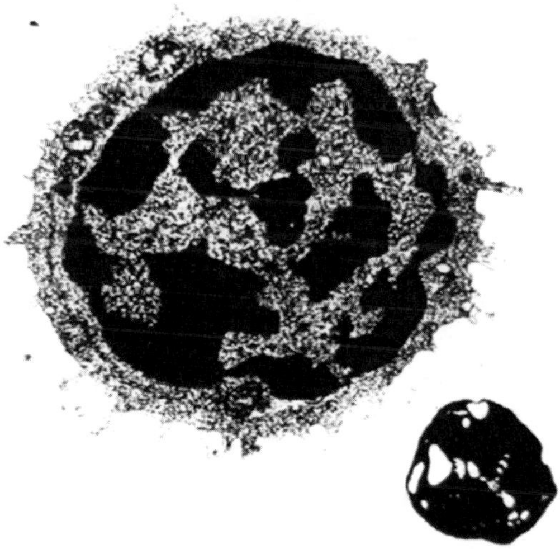

Figure 6. Electron microscope image of mature lymphocytes T (D = 1.20).

Now, the particles present in respirable dust with an aerodynamic diameter greater than 5 μm are considered harmful to health since, by depositing due to their weight, they are first imprisoned in a layer of mucus to protect the hair cells and respiratory bronchioles and then re-expelled through the upper parts of the lung to the rhythmic movement of the cilia (it is noted that nicotine has a paralyzing effect on ciliary movement so the effectiveness of removing dirt is reduced). On the other hand, minute particles of respirable dust with a diameter of less than 5 µm are deposited mainly along the alveolar walls without hair cells, thus evading the first pulmonary defensive barrier. The deposition kinetics in the alveoli, therefore, depends on the interactive surface and the irregularity of the particle shape. This means that the particle's fractal structure affects the deposition rate and configures the extent of damage inflicted on lung tissue.

These particles with a strong surface not only bind with affinity the carcinogens present in smoke; but activate them, making them much more aggressive towards the bronchial tissue than they would have been in the absence of dust particles with a fractal structure. The previous considerations on the shape of the particles also apply to respirable dust. However, in certain situations, the fractal structure of the dust grains may indicate a lower risk of lung penetration. By way of example, the state of silicosis generated by the inhalation of silicon dust is worth mentioning, the physical rather than chemical state of which is decisive and in which the greatest danger comes from the inhalation of crystalline forms of SiO_2 rather than particles with an amorphous or fractal structure.

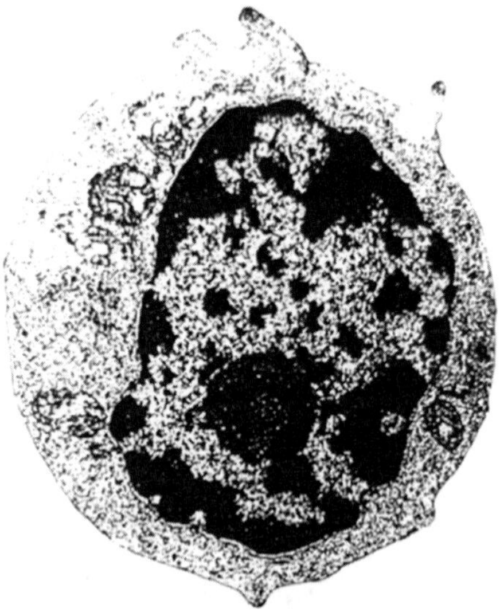

Figure 7. Electron microscope image of acute lymphoblastic leukemia (D = 1.11).

Due to their fractal properties, amorphous powders have fluidizing properties, so powder with a high fractal structure obtained by the combustion of silicon tetrachloride in an oxygen and hydrogen flame (fumed silica) is added to the food powders. The addition of these thinning agents seems to make food powders more soluble and medium metabolizable by the body.

Conclusion

Among the many applications of fractal analysis, it is helpful to mention some inherent to human tumor cell biology (Nonnenmacher, 1989), such as the determination of the surface of membrane proteins, the evaluation of the degree of malignancy of prostate cancer (Irinopoulou et al., 1993), the kinetics of ionic transport channels (Nonnenmacher, 1994), the determination of the degree of irregularity of the interface area between the epithelial tissue and the connective tissue of the oral cavity (Landini, & Rippin, 1993), the kinetic properties of membrane lipids (Nonnenmacher et al., 1990), the process of structural and functional differentiation of the optic nerve (Smith et al., 1994), the measurement of the degree of differentiation of neurons of the central nervous system (Smith et al., 1989), the effect of viscosity on the fractal process of neural arborization (Caserta et al., 1992), and the evaluation of tumor areas in sonographic findings of the breast (Caldwell et al., 1989). Fractal organization is even proposed as a design principle in living organisms (Weibel, 1991).

Project n. 31-25702.88, funded by the Swiss National Science Foundation.

References

Baumann, G., Barth, A., & Nonnenmacher, T. F. (1994). Measuring fractal dimensions of cell contours: practical approaches and their limitations. In Nonnenmahcer, T. F., Losa, G. A., Weibel, E. R. (eds) *Fractals in biology and medicine*, Basel: Birkhäuser Verlag, 182-189.

Caldwell, C. B., Stapleton, S. J., Holdsworth, D. W., Jong, R., Weiser, W., Cooke, G., & Yaffe, M. J. (1989, May). Characterization of mammographic parenchymal pattern by fractal dimension. In *Medical Imaging III: Image Processing* (Vol. 1092, pp. 10–16).

Caserta, F., Hausman, R. E., Eldred, W. D., Kimmel, C., & Stanley, H. E. (1992). Effect of viscosity on neurite outgrowth and fractal dimension. *Neuroscience Letters, 136*(2), 198-202.

Irinopoulou, T., Rigaut, J. P., & Benson, M. C. (1993). Toward objective prognostic grading of prostatic carcinoma using image analysis. *Analytical and quantitative cytology and histology*, *15*(5), 341-344.

Kaye, B. H. (1989). Fine particle look-alikes. *A random walk through fractal dimensions*. VCH Verlagsgesellschaft mbh, 117.

Kaye, B. H. (1989). Fractal structure of some respirable dusts. *A random walk through fractal dimensions*. VCH Verlagsgesellschaft mbh, 95.

Landini, G., & Rippin, J. W. (1993). Fractal dimensions of the epithelial-connective tissue interfaces in premalignant and malignant epithelial lesions of the floor of the mouth. *Analytical and quantitative cytology and histology*, *15*(2), 144–149.

Losa, G. A. (1994). Fractal properties of pericellular membrane from lymphocytes and leukemic cells. In Nonnenmacher, T. F., Losa, G. A., Weibel, E. R. (eds) *Fractals in biology and medicine* (pp. 190–200). Basel: Birkhäuser.

Losa, G. A., Baumann, G., & Nonnenmacher, T. F. (1992). Fractal dimension of pericellular membranes in human lymphocytes and lymphoblastic leukemia cells. *Pathology-Research and Practice*, *188*(4-5), 680-686.

Loud, A. V. (1968). A quantitative stereological description of the ultrastructure of normal rat liver parenchymal cells. *The Journal of Cell Biology*, *37*(1), 27–46.

MacAulay, C., and B. Palcic. "Fractal texture features based on optical density surface area. Use in image analysis of cervical cells." *Analytical and quantitative cytology and histology* 12.6 (1990): 394-398.

Mandelbrot, B. B. (1983). The Fractal Geometry of the nature. Freeman, San Francisco.

Nonnenmacher, T. F. (1989). Fractal scaling mechanisms in biomembranes: oscillations in the lateral diffusion coefficient. *European Biophysics Journal*, *16*, 375-379.

Nonnenmacher, T. F. (1994). Spatial and temporal fractal patterns in cell and molecular biology. In Nonnenmacher, T. F., Losa, G. A., Weibel, E. R. (eds) *Fractals in Biology and Medicine* (pp. 22-37). Basel: Birkhäuser Basel.

Nonnenmacher, T. F., Baumann, G., Barth, A., & Losa, G. A. (1994). Digital image analysis of self-similar cell profiles. *International journal of bio-medical computing*, *37*(2), 131-138.

Nonnenmacher, T. F., Baumann, G., & Losa, G. A. (1990). Self-organization and fractal scaling patterns in biological systems. *Trends in Biological Cybernetics. Trivandrum India: Publication Manager, Research Trends, Council of Scientific Research Integration*, 65-73.

Paumgartner, D., Losa, G., & Weibel, E. R. (1981). Resolution effect on the stereological estimation of surface and volume and its interpretation in terms of fractal dimensions. *Journal of microscopy*, *121*(1), 51-63.

Reith, A. (1973). The influence of triiodothyronine and riboflavin deficiency on the rat liver with special reference to mitochondria. A morphologic, morphometric, and cytochemical study by electron microscopy. *Lab Invest* 29, 216–227.

Rigaut, J. P. (1991). Natural objects show fractal grey tone functions: a novel approach to automated image segmentation, using mathematical morphology. *Acta Stereol* 6, 799-810.

Smith Jr, T. G., & Behar, T. N. (1994). Comparative fractal analysis of cultured glia derived from optic nerve and brain demonstrate different rates of morphological differentiation. *Brain Research, 634*(2), 181-190.

Smith Jr, T. G., Marks, W. B., Lange, G. D., Sheriff Jr, W. H., & Neale, E. A. (1989). A fractal analysis of cell images. *Journal of Neuroscience Methods*, *27*(2), 173–180.

Weibel, E. R., Stäubli, W., Gnägi, H. R., & Hess, F. A. (1969). Correlated morphometric and biochemical studies on the liver cell: I. Morphometric model, stereologic methods, and normal morphometric data for rat liver. *The Journal of Cell Biology*, *42*(1), 68-91.

Weibel, E. R. (1991). Fractal geometry: a design principle for living organisms. *American Journal of Physiology-Lung Cellular and Molecular Physiology*, *261*(6), L361-L369.

Chapter 2

Computational Algorithms to Study the Chaotic Dynamics of the Logistic Equation

Javier Montenegro Joo*

Facultad de Física, Universidad Nacional Mayor de San Marcos, UNMSM, Lima, Perú
VirtualDynamics: Science and Engineering Virtual Labs, Lima, Peru

Abstract

Understanding the behavior of the Logistic Equation is critical for those taking their first steps in the study of Chaos theory, and the best way to get familiar with chaotic dynamics is by developing computer programs to simulate it; with this in mind, this work deals mainly with the computational algorithms to study the dynamics of the mentioned equation. Implementing the computer programs with the algorithms in this work allows visualizing the dependence on initial conditions of the logistic equation, a feature typical of chaotic systems; this is made evident by running the logistic equation starting from slightly different initial conditions and may be graphically shown under slow motion on the computer screen. A special case of dependence on initial conditions is the Butterfly effect, which may also be visualized with the algorithms in this chapter. The reported algorithms also allow experimentally assessing up to five values of Feigenbaum's delta by plotting the equation's bifurcations cascade. This work also deals with the parabolic-shaped Maps of Return of the logistic equation.

Keywords: chaos, bifurcation cascade, butterfly effect, Feigenbaum's Delta, iterative process, logistic equation, return map

---

* Corresponding Author's Email: jmontenegroj@unmsm.edu.pe

In: Complexity and Fractality in Human Change and Beyond
Editor: Nicoletta Sala
ISBN: 979-8-89530-575-1

1. Introduction. Genesis of the Logistic Equation

The origin of the logistic equation can be traced back to the mid-19th century, when the Belgian mathematician Pierre-François Verhulst (Verhulst, 1838) developed a model to describe the growth of a population over time. This model, known as the logistic equation or Verhulst's equation, is used in biology, ecology, and other sciences to model the population growth of organisms since it meets the required conditions for a population growth model to approach reality. Verhulst was inspired by the work of Thomas Malthus, who had proposed exponential growth models for populations. However, Malthus's model did not account for resource limitations, which led Verhulst to introduce the logistic growth equation as a more realistic alternative. Verhulst introduced the logistic equation as a modification of the exponential growth paradigm, incorporating a carrying capacity to represent limits in population growth in a given environment. The Verhulst equation describes how a population approaches and stabilizes around its maximum. This representation better reflects real-world scenarios where resources are finite.

The Malthusian exponential growth model represents unrestricted growth, where a population continuously increases at a fixed percentage rate. In contrast, through the so-called carrying capacity, the Verhulst logistic equation considers limiting factors, which limit population as resources become scarce.

Exponential growth is unbounded, whereas logistic growth decreases over time due to resource constraints.

2. The Standard Logistic Function

A generic or standard logistic function or simply the logistic curve is a familiar s-shaped curve, usually known as a sigmoid curve, whose mathematical expression is

$$f(x) = \frac{L}{1+e^{-k(x-x_0)}} \begin{cases} -\infty < x < +\infty \\ f(x) \to L, x \to +\infty \\ f(x) \to 0, x \to -\infty \end{cases}$$

where x_o is the x value of the mid-point of the function f(x), L is the supremum of the values of the function, and k is the logistic growth rate or the steepness of the curve. Before studying non-linear dynamics and chaos, this author became acquainted with this function for L = 1 when he dealt with multilayered perceptrons, a type of neural network. Figure 1 shows the graph of the logistic function for four different values of k; the higher the value of k, the steeper the curve, whose minimum and maximum values are zero and L, respectively.

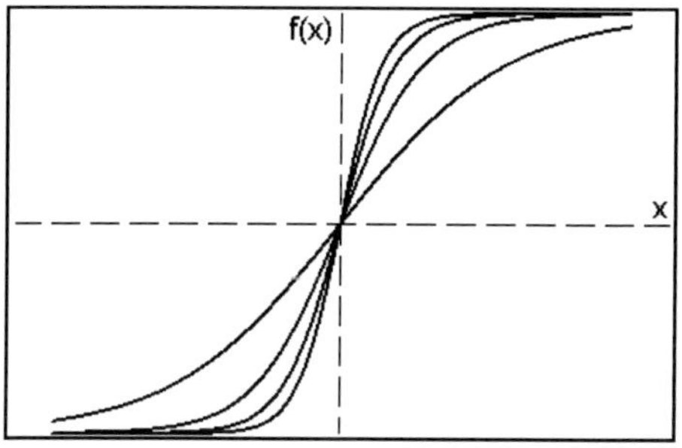

Figure 1. The generic or standard logistic function for four different values of k.

3. The Verhulst Logistic Equation

The Verhulst Logistic Equation is the mathematical representation; this is a mathematical model of the evolution of a population as time elapses, and this equation takes into account environmental limitations and the finiteness of resources. As environmental limitations, factors like available space, weather conditions, and the presence of predators must be considered. Finiteness of resources refers to the existence of water and food. According to Verhulst, the number of individuals in a population or simply the population at a given time t is given by

$$p(t) = \frac{k}{1 + \left(\frac{k - p_o}{p_o}\right) e^{-rt}}$$

In this equation, $p(t)$ is the population at time t. Some authors write N instead of p, and, in those cases, N refers to the number of individuals in the population, k is the carrying capacity of the environment, this is the maximum number of individuals that the environment can sustain according to the restrictions and existing resources, the initial population is represented by p_o and r is the growth rate. The growth rate of a population is the percentage change in the number of individuals in a population over a specific period, which may be a year, a generation, etc. Notice that the growth rate may be positive if the number of new-born individuals overcomes the number of those passing away, and it will be negative in the opposite situation. Obviously, the development of the logistic equation rests on the assumption that the population growth rate is proportional to the current population and that there is maximum capacity in the environment. To deduce an expression representing the change in the population as time goes by, the first temporal derivative of the previous expression must be taken:

$$\frac{dp}{dt} = \frac{-kre^{-rt}\left(\frac{p_o-k}{p_o}\right)}{\left[1+e^{-rt}\left(\frac{k-p_o}{p_o}\right)\right]^2}$$

Then, after replacing it with the original expression for p(t), this expression turns.

$$\frac{dp}{dt} = \frac{-1}{k}\left[p^2re^{-rt}\left(\frac{p_o-k}{p_o}\right)\right]$$

$$\frac{dp}{dt} = rp\left(1-\frac{p}{k}\right)$$

This last expression is a differential equation describing the change in the population as time goes by; it considers the carrying capacity of the environment k and the growth rate r.

Note that as the population p increases, the value of (1- p/k) decreases, thus reducing the growth change, and when the population reaches the carrying capacity k, the growth change turns zero, indicating that the population stabilizes. In other words, when the growth rate r is high and the carrying capacity k is low, exponential population growth is observed at first. However, as the population approaches the carrying capacity, growth slows and stabilizes at a constant value.

On the other hand, when the carrying capacity k is high, and the growth rate r is low, a slow and stable population growth is observed from the beginning. The population quickly stabilizes at a value close to carrying capacity. When the growth rate r is equal to zero, the population remains constant at a value equal to the carrying capacity, which is why an equilibrium exists. The logistic equation has been widely applied in various areas of study, such as epidemiology, species conservation, natural resource management, and economics. It is used to predict and get acquainted with changes in the population of organisms in response to various factors, such as the availability of resources, interspecies interaction, and conservation efforts.

One of the most well-known applications of the logistic equation is the study of the dynamics of human populations (May, 1976). Throughout history, the evolution of the human population has been the subject of great interest and concern. The logistic equation has proven to be a valuable tool in understanding and predicting the human population and assessing the impact of different control strategies and family planning policies.

Returning to the previous equation, dividing both sides of it by k, and replacing p/k by x, the well-known form of the logistic equation is finally obtained:

$$\frac{d}{dt}\left(\frac{p}{k}\right) = r\frac{p}{k}\left(1 - \frac{p}{k}\right)$$

$$\frac{dx}{dt} = rx(1 - x)$$

This nonlinear differential equation describes how the population changes as time evolves. In practice, when trying to make a graph of this equation, the temporal derivative on the left side is replaced simply by y(x), and the growth rate r is replaced by A.

4. Studying the Logistic Equation Using Computer Simulations

The Logistic Equation, the model of population evolution engaged during a long time as the primary reference for studies of chaotic dynamics, is given by (Roderick, 1987; Feigenbaum, 1980; Peitgen, 1991; Hilborn, 2001).

$$y(x) = Ax(1 - x) = A(x - x^2) \tag{1}$$

and as is evident in Figure 2, the graphic of y *versus* x is a parabola whose maximum height depends on the parameter 'A,' known as the Linear Growth Rate.

The flowchart in Figure 2 depicts the computational procedure to graph the parabolas sketched in the exact figure. The process starts by setting the value of the parameter 'A,' the initial value of 'x,' and the value of the increments Δx used in the computations, the variable 'k' is a counter that reports the number of steps in the calculations. The initial value of 'x' should be a small positive number less than 1.

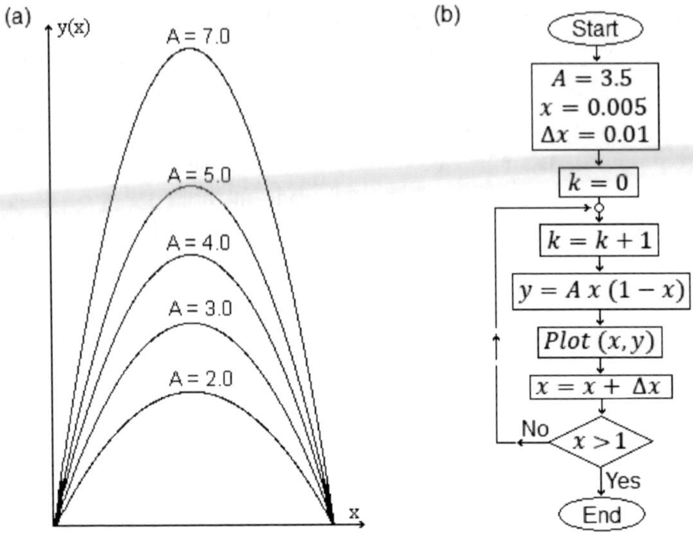

Figure 2. (a) Parabolas resulting from plotting the Logistic Equation for several values of the Linear Growth Rate 'A.' The larger the value of 'A,' the higher the altitude of the parabola. (b) Flowchart of the computational algorithm to generate the parabolas.

The algorithm consists of an iteration (a loop) where, after each assessment of the logistic equation, the plotting of (x, y) pairs may be made on the computer screen, or these pairs may be stored and used to make the plotting when the program finishes its computations, when $x = 1$. During the process, after computing each (x, y) pair, the value of 'x' is incremented by adding to it the predefined Δx and the loop is repeated.

In the flowchart in Figure 2, the initial value of x is 0.005, and the iteration process continues until x > 1.

5. Iterative Version of the Logistic Equation

When studying the number of individuals (the population) as a function of generation, the logistic equation is expressed as

$$y = x\,(y - y^2) = x\,y\,(1 - y) \qquad (2)$$

In this equation, 'x' is the increasing number of generations in a particular group of individuals, and 'y' is the number of individuals –which depends on the available resources- in each generation. In this case, the logistic equation is an iterative equation where the population 'y' in a given generation 'x' is a function of the population in the previous one; this is a mathematical function represented as $y_n = F(y_{n-1})$.

Figure 3 shows a graph of population versus generation in the Logistic equation, on the horizontal axis appears the generations $(1.0 \leq x \leq 4.0)$ whereas the vertical axis shows the number of individuals in those generations. As can be seen, the initial population is very small and gradually increases until finally the available food is no longer sufficient. As a result, some individuals die, and their numbers begin to fluctuate from generation to generation, first between two sets, then between four sets, and so on, until it becomes impossible to predict the number of individuals in the next generation.

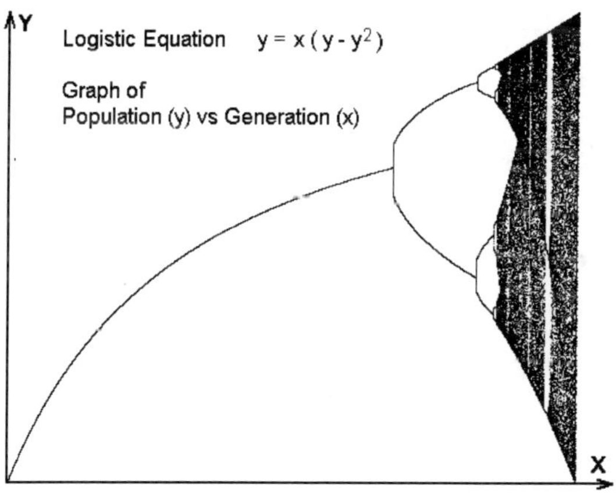

Figure 3. Dynamics of a population whose evolution is modeled by the logistic equation.

6. Interpretation of the Logistic Equation

Even though the logistic equation is not considered a good population dynamic model, it is good enough as a first approximation to an evolutionary model.

Consider a barn where an amount more or less constant of grains falls off the bags containing them permanently. Sooner or later, some organisms like insects or rodents who feed from these grains will appear, and if initially, their number is small, they will cheerily multiply, and their population will increase generation after generation in a way proportional to the number of individuals (see Figure 3). The larger the number of lodgers, the higher their population for each new generation; a day will come when the number of individuals in the population is so large that the number of grains will not be enough to feed all of them, and some of them will starve. The next generation, those lodgers surviving the scarcity, will reproduce again, and all of them survive with the available grains. However, in the subsequent generation, there is an overpopulation, and many dwellers die of hunger. In this way, the number of individuals has begun to alternate from generation to generation between two numbers; this is why the population has experienced a bifurcation. The process continues, and new bifurcations in the population appear. Eventually, the system evolves towards a state where the population alternates between increments and decrements in the number of individuals, making it impossible to foresee their number in the next generation. This means that the system is experiencing a chaotic process.

7. Dynamics of the Logistic Equation

In the graph of the logistic equation (Figure 3), each generation in which the number of individuals is equal to or very close to an earlier one, a period is completed, which is why this cascade of population bifurcation is also known as a cascade of period bifurcation.

It can be seen in Figure 3 that at the beginning, when the population only increases, there is only a single period during a long time; notice, however, that this period is not constant; it is slowly increasing. Then, after the first bifurcation, there are two periods; after the second bifurcation, there are four periods, and this dynamic continues until, eventually, it becomes impossible to foresee what the following number of individuals (population) will be, and the system enters chaos. Figure 4 shows the transition to chaos in the logistic equation; this image used to be the benchmark in early chaos studies. Notice that Figure 4 expressly displays the bifurcation cascades appearing in Figure 3.

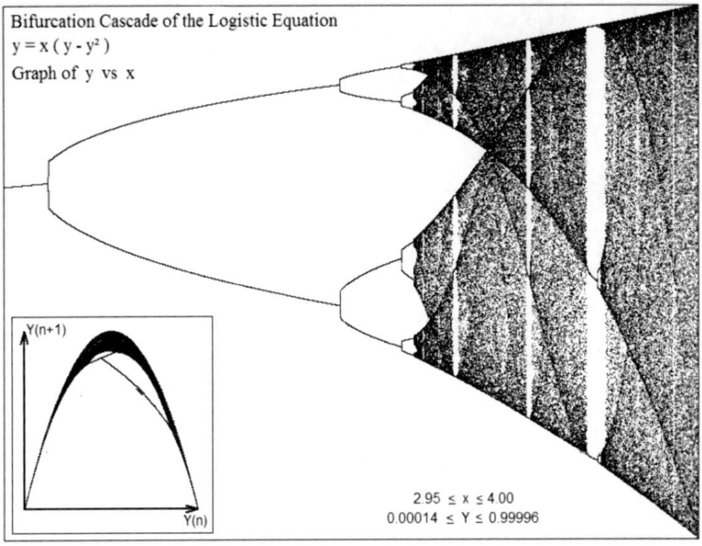

Figure 4. A cascade of period bifurcations in the logistic equation. The 'x' variable, indicating generation, is along the horizontal and goes from 2.95 through 4.00, the 'y' variable -indicating population- is on the vertical and goes from 0.00014 through 0.99996. The inset shows the associated Return Map, which is a parabola.

A bifurcation cascade is a mathematical function of the form $y = y(x)$, and as a function, each value of 'x' must correspond to a single value of 'y,' however, the cascade gives the impression that there are two or more values of 'y' for some values of 'x.' Actually, after the first bifurcation, the values of 'y' alternate between two values as 'x' increases; it is not that there are two values of 'y' for each 'x.'

Then, at the second bifurcation, each 'y' of the first bifurcation alternates between two other values, and the population is now alternating among four numbers; this process continues until, eventually, the system enters blatant chaos, where the period becomes infinite and therefore the 'y' amounts of individuals in the population are practically unpredictable.

8. Iterative Process

An iterative process with an equation may be considered a feedback process that starts by assessing an arbitrary value 'x_o' in the equation and reinserting the output 'y' as a new input at each iteration.

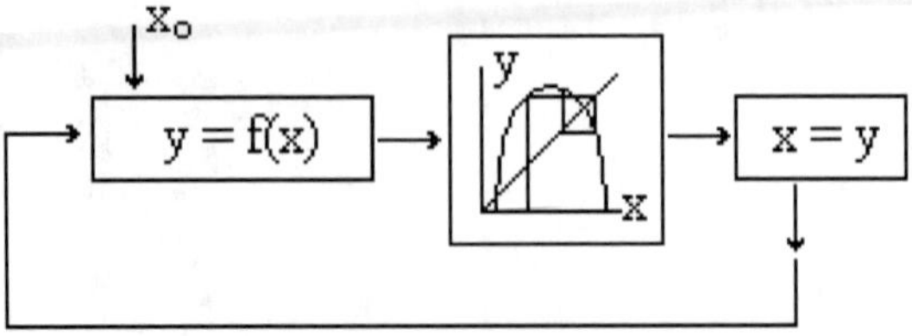

Figure 5. Iteration process. In this sketch, the graph of y versus x is not part of the process; it is there just to visualize the effect of computing the output Y as a function of the input X.

The dynamics of the logistic equation are based on iterations, and Figure 5 is the graphical representation of an iterative process, also known simply as an iteration; the process starts with an arbitrary value 'x_o' being inserted in an equation y = f(x), the output of this computation is 'y' and, this value is reinserted as 'x' in the equation, then the new output is reinserted in the equation and so on. Usually, the iteration process involves a relatively large number of inputs and outputs, and obviously, the larger the number of cycles, the better the results.

In Figure 5, a plotting of y vs x is made after each iteration to visualize the relationship between input and output. Sometimes, the values of 'x' and 'y' are stored in an array after each computation, and the plotting of y vs x or the analysis of resulting data is made at the end of the whole iterative process.

9. Iterations in the Logistic Equation

In the Logistic Equation given by equation (1), $y(x)$ is the output corresponding to an input x. Since, at each iteration, the previous output becomes the new input, the iterative version of this equation may be expressed as

$$x_{n+1} = Ax_n[1 - x_n], n = 0,1,2, \ldots, N \tag{3}$$

In this equation, 'x' is the number of individuals (the population) in a given generation. In the iterations performed with the logistic equation, the first input, the initial value, is an arbitrarily chosen value x_o, this is inserted in the equation, which means that the equation is calculated with this first value. The resulting value, the output of this first computation, is used as the second input, and the output is used as the third input, and so on. Geometrically, this process involving the input of the preceding output is represented with the help of a line at 45° in the graph of y versus x (Figure 6).

To construct a y(x) function for the logistic equation, change x to y and A to x in the preceding equation:

$$y_{n+1} = x\,y_n[1 - y_n] \rightarrow y = xy\,(1 - y) \tag{4}$$

This equation has been used in this report to generate all the graphs where the logistic parabola (whose shape depends on the linear growth rate (A) and a diagonal at 45° appear. To initiate the computations, the value of A and a starting value x_o must be set by the user of the computer program.

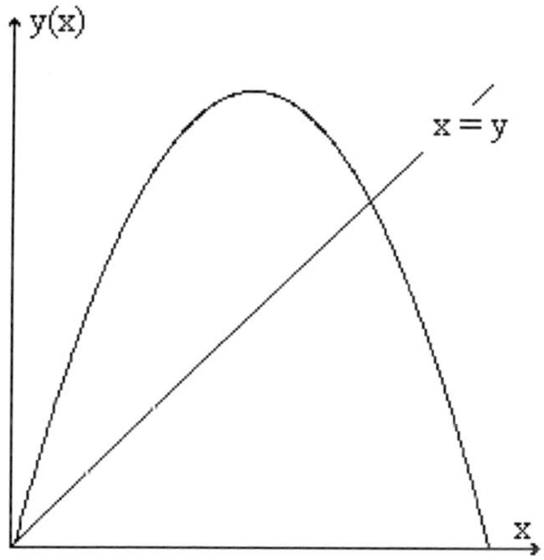

Figure 6. The logistic parabola for A = 4.00, and the diagonal line at 45°, whose equation is x = y. All points along the diagonal have equal values of x and y. The use of this diagonal plays an important role when visualizing the iterations with the Logistic Equation.

The process of inputting the previous output is repeated until a predefined number of iterations is completed (maybe 500) or some other condition set by the computer programmer is fulfilled. In the case of the iteration graphs shown here, the initial value x_o is saved and compared with each iteration output, the computation is stopped when the output results equal to the stored value x_o , or when there is no change in successive outputs.

Figure 7 shows the dynamics of the first five iterations of equation (3), starting with $x_o = 0.65$, which is the first input; this value generates an output inserted in the equation as the second input, and its output is once again inserted in the equation, and so on. In Figure 7, the vertical arrows represent inputs to the logistic equation, whereas the horizontal arrows are the associated outputs, and the line x = y at 45° represents the transformation of a previous output into a new input. When depicting the dynamics shown in Figure 7, x(n + 1) is along the vertical, and x(n) is on the horizontal.

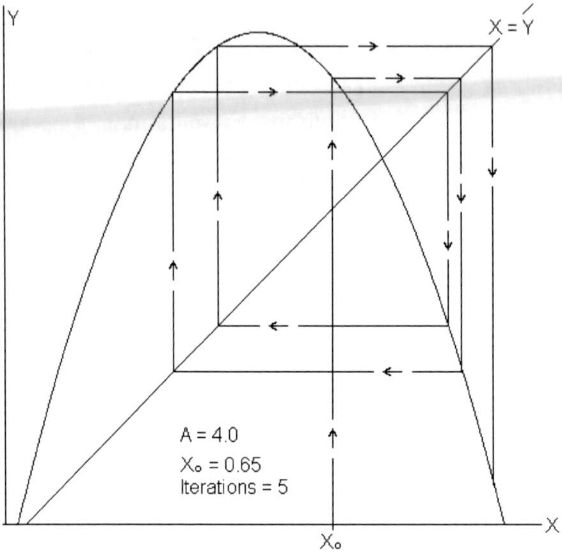

Figure 7. Parabola of the Logistic Equation for A = 4.0. The x = y line at 45° is shown. This line is used to convert a previous output value into a new input. The arrows placed on the lines indicate the sequence of inputs and outputs and the fact that, at any moment, an output becomes the new input.

10. Effect of the Linear Growth Rate in the Dynamics of the Logistic Equation

It is experimentally observed (see Figure 8) that when $A \leq 1$, all initial values x_o generate sequences of outputs tending to zero ($x \to 0$), which is interpreted as the extinction of the population.

When the values of 'A' are close to 1 or 3, all initial populations x_o are attracted towards some fixed value, meaning that the number of individuals in the population stabilizes in some number (see Figure 9). When 'A' is around 2, all initial conditions x_o are attracted towards a fixed point, the same output is obtained after a few iterations, which means that the population stabilizes.

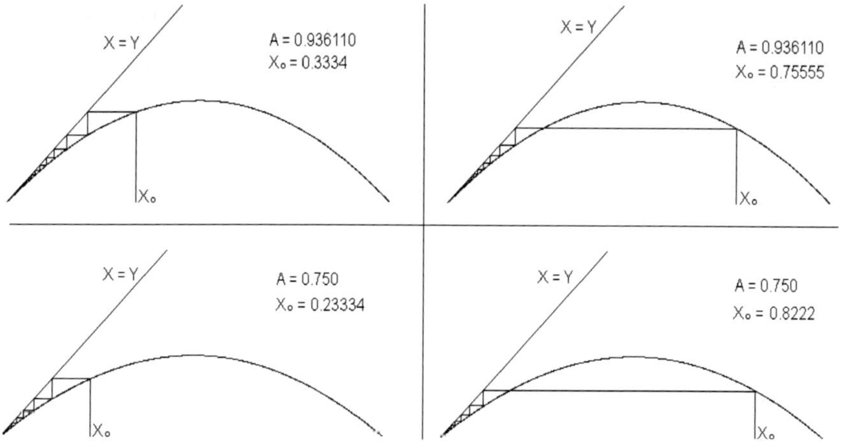

Figure 8. Two pairs of examples of the situation when any initial population is extinguished. The top and the bottom images show two cases for the same value of the linear growth rate 'A.' As it is evident, any initial value x_o of the population evolves towards extinction when $A \leq 1$.

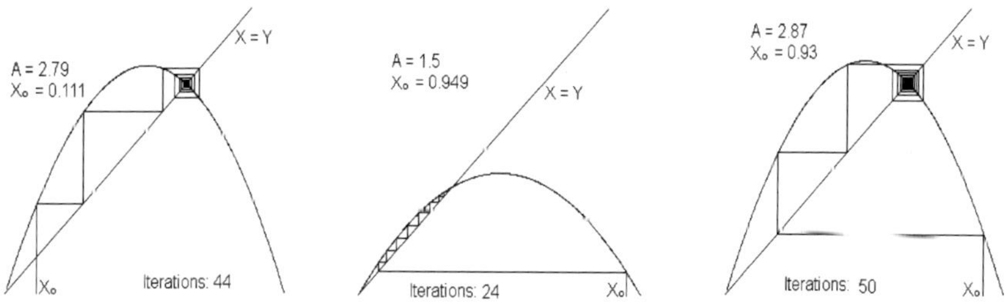

Figure 9. Three examples of the case with extreme values of 'A' between 1 and 3 show that whatever the initial population, it inexorably stabilizes in some fixed value. In each case, the data associated with each image and the number of iterations after which the iteration process was automatically stopped are shown.

Figure 10 shows the situation when $A \geq 3$, in these cases, for any initial value x_o of the population, the number of individuals x in the successive populations alternates between two values, which are more distant as 'A' is increased, until around A = 3.75 when some glimpses of chaotic behavior start to be evident (Figure 11).

Finally (see Figure 11), when A = 4.0, the population takes completely unexpected values, making it impossible to foresee the number of individuals in the population; this chaos has taken control of the dynamics.

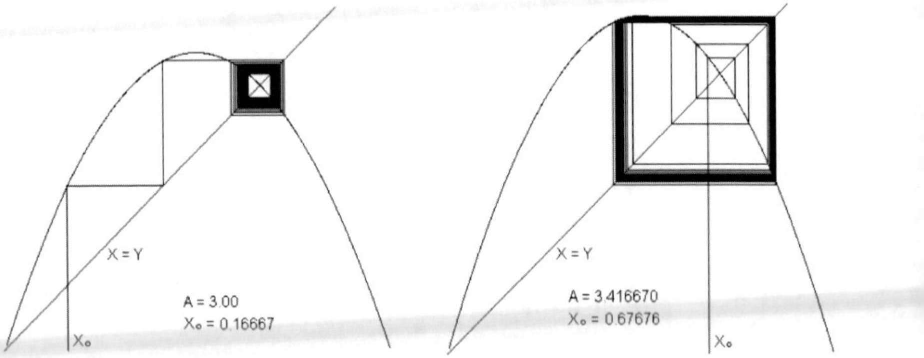

Figure 10. Two case examples with A ≥ 3.0.

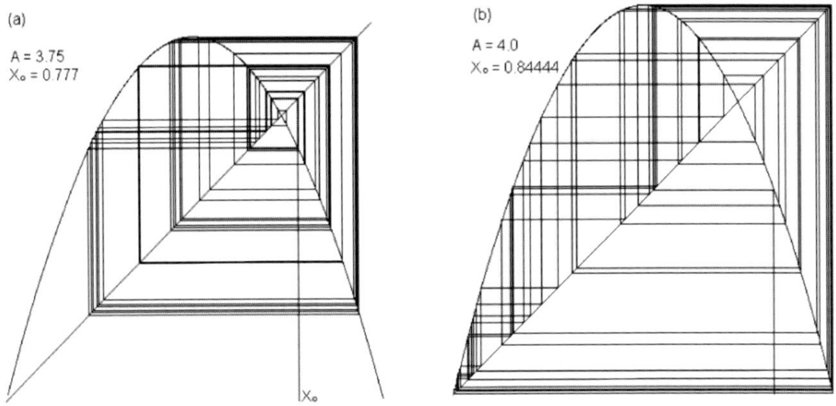

Figure 11. (a) When A is around 3.75, some glimpses of chaotic behavior start to appear (b) Chaos is declared when A = 4.

11. The Butterfly Effect in the Logistic Equation

The butterfly effect (Lorentz, 1993) in chaotic systems establishes that a very small change in the initial conditions of a dynamic system generates a huge difference in another place. In order to appreciate the first example of the Butterfly effect in the Logistic equation, observe and compare what occurs in Figures 12 and 13. Figure 12 depicts the dynamics of the iterations with the logistic equation for $A = 4.0$ and an initial value of $x_o = 0.250000$. Since the computer program developed by this author has been prepared to automatically halt as soon as the output gets repeated, in the case of Figure 12, the process was automatically stopped after eight iterations when the outputs of the iterations started to repeat.

Figure 13 shows two cases where the iterations in the logistic equation started at slightly different positions on both sides, as shown in Figure 12. In the two cases shown, the iterations were respectively started at $x_o = 0.25 - 0.00001 = 0.2499900$ and $x_o = 0.25 + 0.00001 = 0.2500100$. As it can be appreciated, the results in both cases shown in Figure 13 are quite different from those shown in Figure 12. This occurs despite the first input value being extremely small in both cases.

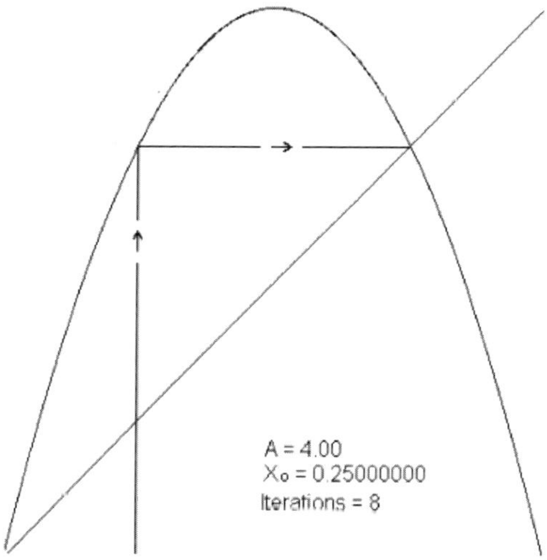

Figure 12. Evolution of the iterations in the Logistic Equation for $A = 4.00$ and $x_0 = 0.250000$ This process was automatically stopped after 8 iterations when the outputs started to repeat.

When comparing Figures 12 and 13, it becomes clear that the results are quite different despite the minimal difference in the initial values. To ease visualization, in both cases in Figure 13, the processes were stopped after 50 iterations; if the process continues, the details become impossible to appreciate. Even after 2500 iterations in the computations of the logistic equation for the two cases displayed in Figure 13, no repeated outputs are detected, but the graphs turn so dense that it is impossible to appreciate any detail.

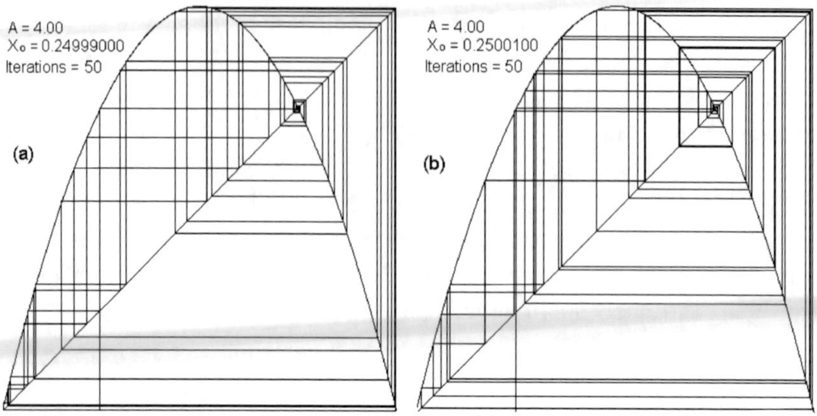

Figure 13. In these two cases, the iteration process started at a minimal distance from the case shown in Figure 12, and, as can be seen, the dynamics are quite different. These two processes were stopped after 50 iterations to avoid a high density of lines.

Figure 14 shows another example of the Butterfly effect in the Logistic Equation. In (a), the initial value is $x_o = 0.75000$, and after 4 iterations, the output value begins to be equal to the input value, which means that the process is not viable and it is automatically stopped. In (b), the process is started at $x_o = 0.75 - 0.00001 = 0.74999$, and as it can be seen, the evolution of the system is completely different from that in (a). In (c), the initial input value is $x_o = 0.75 + 0.00001 = 0.75001$, and again, the evolution is very different from that of (a). Figure 14 verifies once again that starting a chaotic system at a very close position to a reference point may lead the evolution to a completely different situation. In the last two cases shown in Figure 14, the computer simulation was stopped after 50 iterations to ease the visualization of its dynamics. In this case, if the process is not stopped, it is observed that after some 2000 iterations, the graph becomes impossible to appreciate due to the high density of lines.

In general, the formal test to declare a dynamical system as chaotic is re-starting it at a very close position to a reference point and observing that the system evolves to a quite different point after some time; this is when the system runs into the Butterfly effect. Thus, if a dynamic system satisfies the Butterfly effect, it is chaotic. Once the system meets the Butterfly effect, the quantification of its chaotic behavior may be assessed using the maximum Lyapunov exponent.

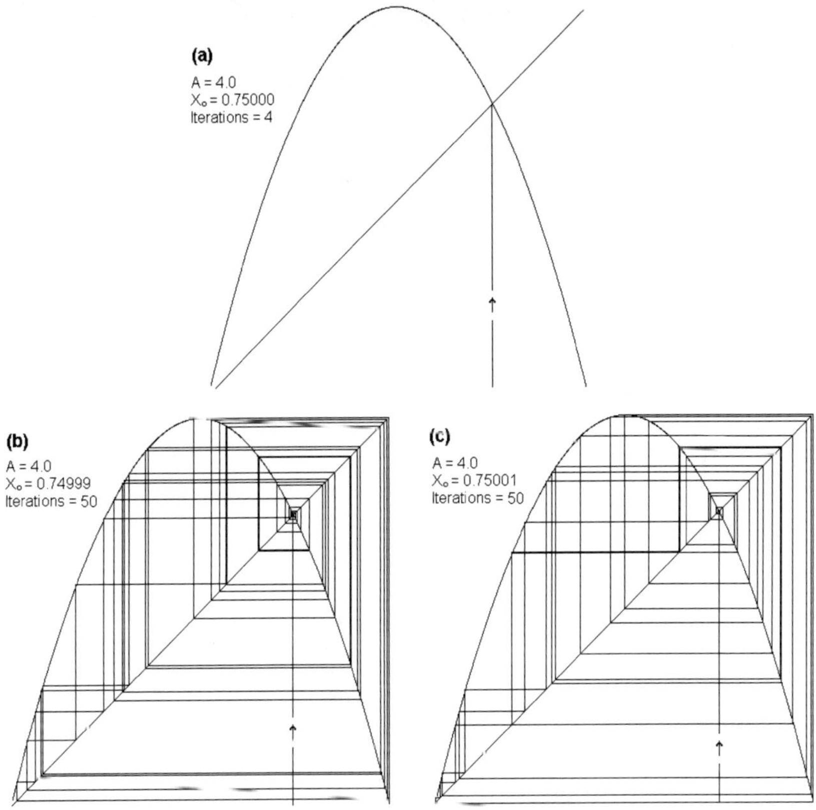

Figure 14. A second example of the Butterfly effect in the Logistic equation.

12. Flowchart of a Computer Program to Graph the Logistic Equation

Figure 15 displays the flowchart of the computer program developed by this author in order to generate the bifurcation cascades of the logistic equation (2) appearing in Figures 3 and 4.

The flowchart shown in Figure 15 allows for generating a graph where the 'x' variable appears on the horizontal, while the 'y' variable appears on the vertical. N_x is the horizontal length of the diagram showing the bifurcation cascades, so it can be the number of pixels the diagram occupies when displayed on the computer screen. If the cascade of bifurcations cannot be generated simultaneously with calculations, the pairs (x, y) can be stored for later graphing with some plotter.

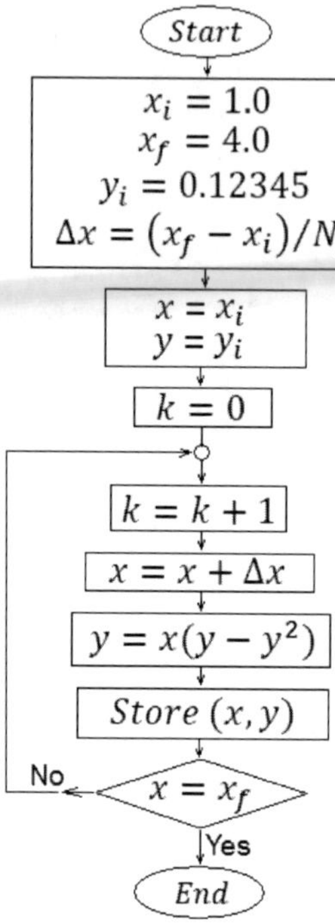

Figure 15. The flowchart of the computer program that generates the bifurcations cascades of the logistic equation (equation 2) is shown in Figures 3 and 4. The logistic equation does not work for x values greater than 4.0.

13. Detection of Bifurcations in the Bifurcation Cascade of the Logistic Equation

The bifurcation cascade of the logistic equation (Figures 3 and 4) shows that the system originally oscillates with a single period. It experiences a bifurcation of the period and oscillates, alternating between two periods; then again, each of these two periods bifurcates, and the system oscillates, alternating among 4 periods. This process continues for 32 periods, after which chaotic behavior is openly declared. This author has identified the values of 'x' in the logistic equation where the bifurcations occur:

- Period 1 to period 2: x = 3.016
- Period 2 to period 4: x = 3.459
- Period 4 to period 8: x = 3.548
- Period 8 to period 16: x = 3.566
- Period 16 to period 32: x = 3.569
- Chaos occurs between x = 3.570 and x = 4.00

To detect the above-mentioned x-values of the bifurcation points, a Virtual Lab that generates a much larger plotting than a computer screen was devised; this graph is rolled up and unrolled by moving the mouse over scroll bars (horizontal and vertical) on the margins of the plotting. Additionally, as the mouse cursor is displaced over the graph on the screen, the Virtual Lab reports the corresponding (x, y) coordinates; in this way, by developing a sufficiently large graph and placing the mouse over the bifurcations, the values of 'x' and 'y' are easily obtained at each point. The same procedure was used to detect where the onset of chaos occurs.

14. The Feigenbaum's Delta

Mitchell Feigenbaum, one of the fathers of Chaos Theory, deduced an equation to measure the proportion of the distances between two contiguous bifurcations in the logistic equation, it is known as the Feigenbaum Delta, and it is given by (Hilborn, 2001):

$$\delta_n = \frac{x_n - x_{n-1}}{x_{n+1} - x_n} = 4.66920161\ldots \tag{5}$$

Each instance of this equation compromises three consecutive bifurcations: x_n is the nth bifurcation, x_{n-1} is the preceding bifurcation, and x_{n+1} is the next. The numerator is the distance between the nth bifurcation and the preceding one, and the denominator is the distance between the nth bifurcation and the next one. With this equation, Feigenbaum reported the same delta for the two first bifurcations as $\delta_2 = \delta_3 = 4.669$ (Feigenbaum, 1980; Peitgen, 1991; Hilborn, 2001). Using the (x, y) detection procedure described above and inserting the values in equation (5), this author has found the following values:

$$\delta_2 = \frac{x_2 - x_1}{x_3 - x_2} = 4.977$$

$$\delta_3 = \frac{x_3 - x_2}{x_4 - x_3} = 4.944$$

$$\delta_4 = \frac{x_4 - x_3}{x_5 - x_4} = 6.0$$

$$\delta_5 = \frac{x_5 - x_4}{x_6 - x_5} = 3.0$$

Note that the first two values are close to those reported by Feigenbaum (Hilborn, 2001).

15. Maps of Return

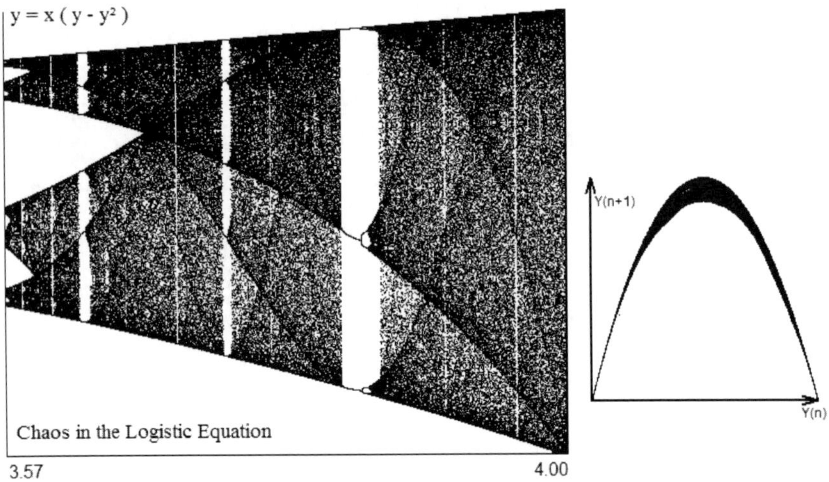

Figure 16. Chaotic event in the logistic equation and the corresponding Return Map.

A map of return is the plotting of a maximum displacement versus the one preceding it, this is Y_{n+1} vs Y_n (Gonzales-Miranda, 2004; Montenegro, 2018; Montenegro, 2019). In Figures 4 and 16, bifurcation cascades with their respective maps of return may be scrutinized. As it can be seen, the map of the return of chaos in the logistic equation is a parabola. In the very early studies of chaos, it was common to think that all chaotic systems should show a parabolic map of return. Now, it is known that this is not necessarily true.

Additionally, it has been experimentally found that the maps of return of chaotic events in physical systems display self-similarity (Montenegro, 2019), and there is strong evidence that they have a fractal structure.

16. Mathematical Zoom of the Bifurcation Cascade

To get insight into the details and the dynamics of the logistic equation, the computer program based on the flowchart shown in Figure 15 may be enabled to make what may be called a 'Mathematical Zoom,' by reproducing the computations associated with a mouse-selected portion of the bifurcation cascade displayed on the computer screen; these are illustrated in Figures 17 and 18. In Figure 17, the highlighted area in the graph of y(x) is illustrated within a rectangle. Figure 18 displays the highlighted region after the corresponding computations have been executed.

Notice in Figure 18 that the development of the highlighted region in Figure 17 seems to render duplicated curves. As explained next, this is an effect of poor graphic resolution that occurs, notwithstanding that the resolution used is the same in both graphs.

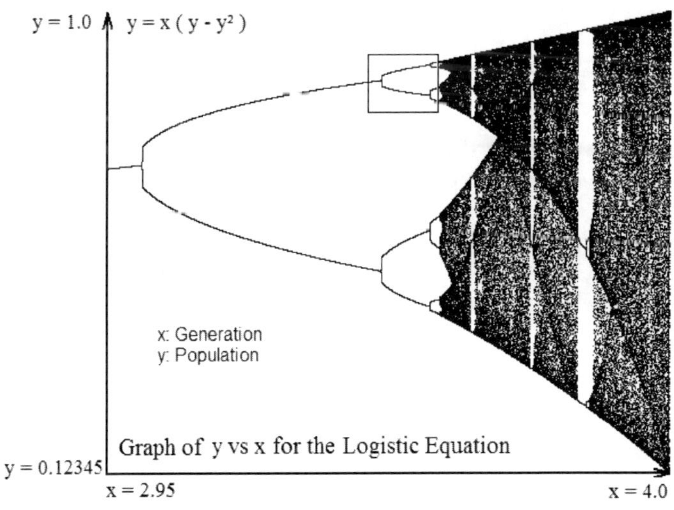

Figure 17. The cascade of bifurcations of the logistic equation. Even though after the first bifurcation, it seems like two values of 'y' for each value of 'x,' the fact is that y(x) is alternating between two values, and a similar effect is evident after the second and third bifurcations. The region selected in the inset is shown in Figure 18.

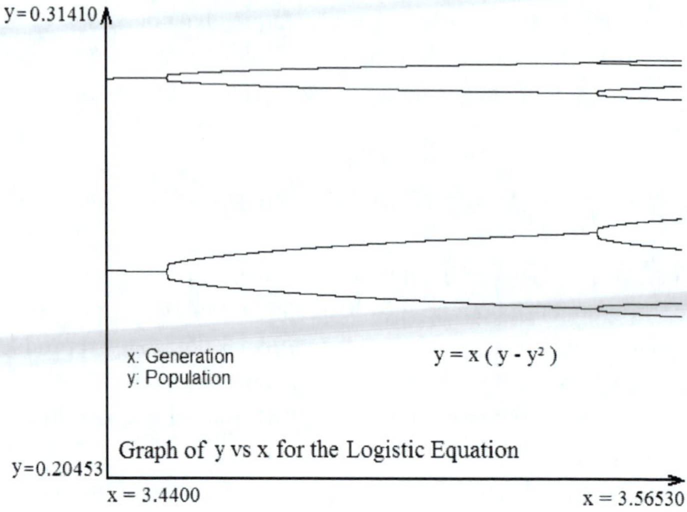

Figure 18. The development of the region is shown in the inset of Figure 17. The same graphic resolution reveals more details when it is applied to a smaller area.

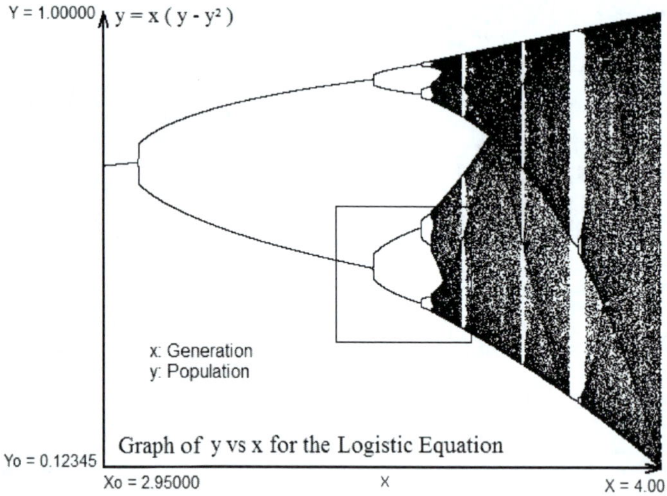

Figure 19. Bifurcation cascade of the logistic equation. The corresponding values of the x and y parameters are shown. The inset selecting the small cascade is shown in Figure 20 after performing the corresponding calculations.

Since Figure 17 embraces a large region, the used graphic resolution comes out to be relatively low, and as a consequence, some details remain hidden; in Figure 18 where the graph is limited to the selected region in the inset in Figure 17 –which is much smaller- the same graphic resolution results higher and hence, better and, the plotting allows displaying some details that remained hidden when the resolution is smaller (Figure 17).

In Figure 18, the values of ‘y’ are alternating between two numerical values very close to one another, and this is visible only with a high graphic resolution; when the resolution is low (Figure 17), both alternating values appear as a single value.

Figures 19 and 20 are shown with the same graphic resolution, and they display another case of making a mathematical zoom of a selected region on the bifurcation cascade.

The period bifurcation cascade displayed in Figure 4 used to be -during a relatively long time- after the beginning of chaos theory, the primary mentioned reference when speaking about chaos. As it can be seen, this cascade may be interpreted as the fact that once chaos starts, the system is doomed to experience chaos forever; it also gives the impression that chaos inexorably increases its intensity once it starts.

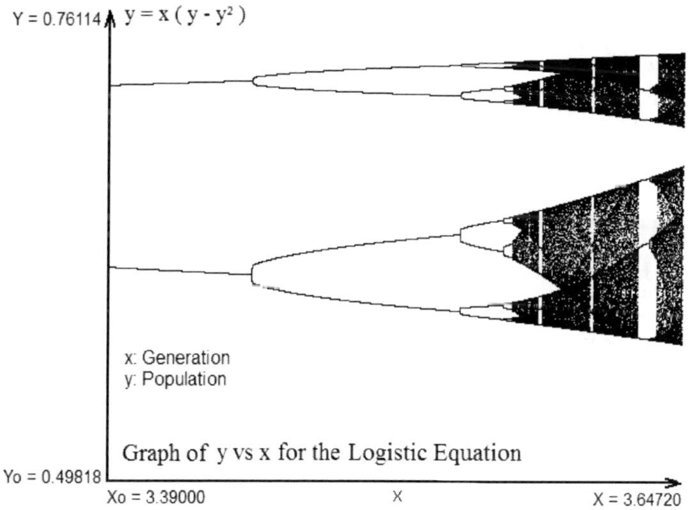

Figure 20. Development of the bifurcation cascade of the selected region is shown in the inset in Figure 19. The same graphic resolution reveals more details when applied to a smaller area.

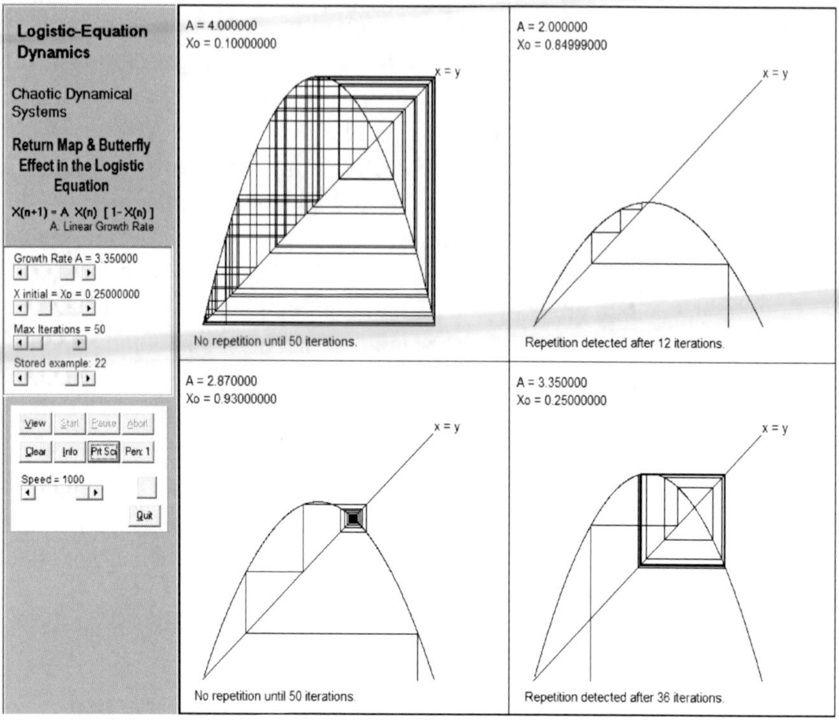

Figure 21. The interface of the computer program to study the dynamics of the Logistic Equation.

By computer-simulation studying the nonlinear damped and forced oscillator, this author found that chaotic events are not only limited in time, having a beginning and an end, but they also are multiple and do not necessarily have the shape shown in Figure 4 (Montenegro, 2014; Montenegro, 2016; Montenegro, 2018).

Figure 21 is a screenshot of one of the virtual labs developed by the author of this work to investigate the dynamics of the logistic equation. In the case shown here, the module is enabled to show four instances of dynamics simultaneously with data entered by the user.

The speed of the plotting development can be set to slow motion mode, which permits watching the evolution of the graphs; the multiple views allow comparison.

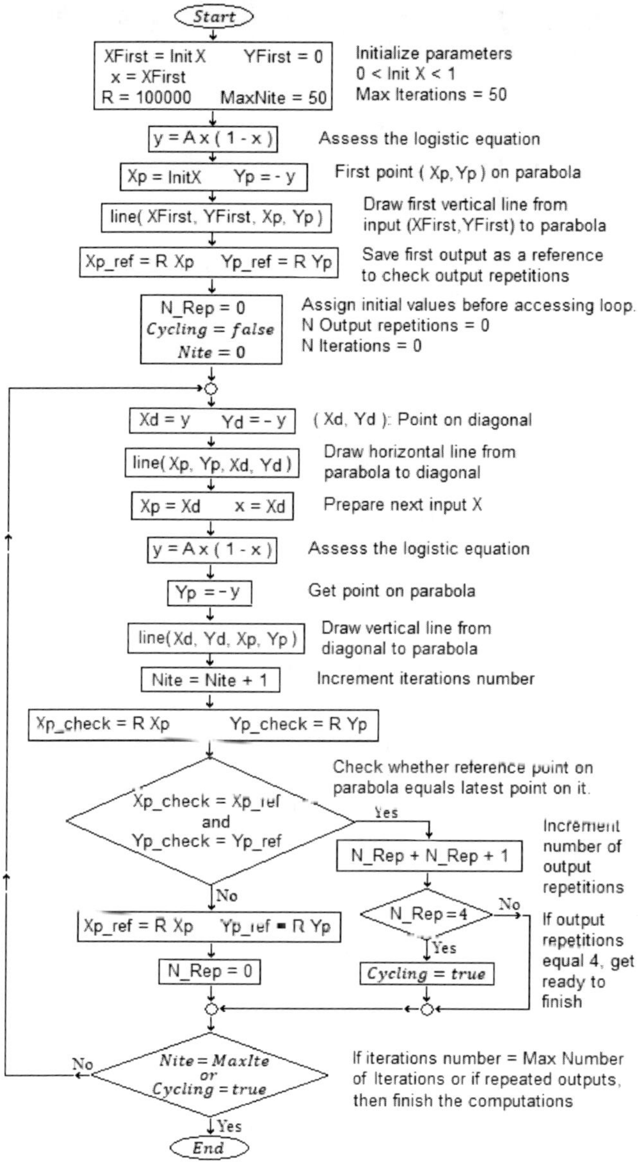

Figure 22. The flowchart of the computer program to generate the dynamics of the logistic equation is shown in Figure 21.

17. Computer Program to Generate the Dynamics of the Logistic Equation

Figure 22 displays the flowchart of the computer program to generate the dynamics shown in Figure 21; this program allows investigation of the behavior of the iterations with the logistic equation.

The program starts by assigning the maximum number of iterations and the initial value of the input parameter x, which must be between 0 and 1. Interested readers may use this flowchart to develop their software using the computer programming language of their preference.

The R-factor in the flowchart is an integer number used to convert a real number into an integer and enable the multiplied parameters to be compared between them.

18. Chaotic Systems and Their Sensitive Dependence on Initial Conditions

A significant characteristic of chaotic systems is their sensitive dependence on initial conditions (Hilborn, 2001; Lorentz, 1993; Riste, 1986), and the logistic equation also displays this feature. The dependence on initial conditions for the logistic equation is visualized in Figures 13 and 14, where an infinitesimal difference in the initial parameter leads the evolution of the system to a very different scenario.

As already mentioned, this dependence is explained by the Butterfly Effect, and it is quantified using the Lyapunov exponent (De Souza-Machado, 1990; Montenegro, 2018; Montenegro, 2022), which results from the data processing of the evolution of the state space of a dynamical system.

Generally, when calculating the Lyapunov exponents during the evolution of a dynamical system, if at least one positive exponent is encountered, it is concluded that the system is chaotic. From the definition of the Lyapunov exponent, it is straightforward to deduce that any expansion of the state space is associated with a positive Lyapunov exponent, whereas any contraction of it implies a negative exponent (Montenegro, 2018) (Montenegro, 2022). There are also hyperchaotic systems (Junhai, 2013; Kapitaniak, 1995); these have more than a single positive Lyapunov exponent, and the average Lyapunov exponent is also a positive number.

Conclusion

This work has presented a concise historical review and the motivation behind the development of the logistic equation by Verhulst, who successfully modified the Malthus equation of exponential growth of populations, making it much more realistic. Some flowcharts of computer programs conceived by this author have been included; these may be used to investigate and become acquainted with the behavior and chaotic dynamics of the logistic equation, which is the most fundamental mathematical model to get immersed in chaos theory. The software based on these flowcharts has allowed the author of this work to visualize the butterfly effect; it has also been possible to detect the coordinates of the bifurcation points in the cascade of bifurcations, and verify some of those obtained by Feigenbaum, one of the fathers of chaos theory. The maps of return are also obtainable from the flowcharts presented in this work.

This work also contains comments and references for researchers interested in the dynamics of the logistic equation, as many of the details that emerge in these dynamics pave the way to understanding the behavior of other more complicated chaotic systems.

References

De Souza-Machado, S, Rollins, R. W., Jacobs, D. T., & Hartman, J. K. (1990). Studying Chaotic Systems Using Microcomputer Simulations and Lyapunov Exponents. *Amer. Journal of Phys*. 58(4), 321–329.

Feigenbaum, M. J. (1980). *Universal behavior in Nonlinear Systems* (69th ed.). Los Alamos.

Gonzalez-Miranda, J. M. (2004). *Synchronization and control of Chaos*, Imperial College Press.

Hilborn, R. C. (2001). *Chaos and Nonlinear Dynamics*, Oxford University Press.

Junhai, M., & Yujing, Y. (2013). Hyperchaos Numerical Simulation and Control in a 4D Hyperchaotic System. Discrete Dynamics in *Nature and Society*, Hindawi Publishing, Article ID 980578

Kapitaniak, T., Thylwe, K. E., Cohen, I., & Wjewoda, J. (1995). Chaos-hyperchaos Transition. *Chaos, Solitons & Fractals*, Vol 5(10), 2003-2011.

Lorenz, E., (1993). *The Essence of Chaos*. University of Washington Press.

May, R. M. (1976). Simple mathematical models with very complicated dynamics. *Nature*, 261(5560), 459–467.

Montenegro, J. J. (2018). *Anatomía del Caos*, Editorial Académica Española.

Montenegro, J. J. (2019). Analyzing Chaotic Dynamics by Means of State Space Tomographies and Dual Maps of Return. *Chaos and Complexity Letters* 13(1), 29-43.

Montenegro, J. J. (2016). Variety and finiteness of Chaotic Events in the NLDF Oscillator. *Chaos and Complexity Letters*, Vol 10-1, 81-87.

Montenegro, J. J. (2014). Multiplicity and Transitoriness of Chaotic Events. *Revista de Investigación de Física*, UNMSM, Vol 17, No 2. (1-6).

Montenegro, J. J. (2018). Tomographies of the State-Space of the Chaotic NLDFO, *Chaos and Complexity Letters* 12(1), 71-82.

Montenegro, J. J. (2022). Evolution of the Lyapunov Exponent During a Chaotic Event in the Nonlinear Damped and Forced Oscillator, *Chaos and Complexity Letters* 16(1), 27-40.

Peitgen, H. O., Jurgens, H., & Saupe, D. (1991). *Fractals for the classroom: Complex systems and Mandelbrot set* (1st ed., Vol. 2). Springer-Verlag.

Pietronero, L., & Tosatti, E. (1985). *Proceedings of the 6th Trieste International Symposium on Fractals in Physics*, North-Holland Physics Publishing.

Riste, T., & Otnes, K. (1986). *The Dimensions of Strange Attractors: A survey of experimental studies. Scaling phenomena in disordered systems*. Plenum Press.

Roderick, J. V. (1987). *Classical Chaos* (69th ed.). American Scientist.

Verhulst, P. F. (1838), Notice sur la loi que la population suit dans son accroissement. *Correspondance Mathématique et Physique*, 10, 113-121.

Wolf A., Swift J. B., & Swinney H. L., Vastano J. A. (1985). Determining Lyapunov Exponents from a Time Series. *Physica* 16D.

Chapter 3

The Numerical Solution of Nonlinear Fredholm Integro-Differential Equations via the Parametric Iteration Method

Mohamadreza Tamamgar*
Department of Mathematics, Islamic Azad University, Gonabad Branch, Gonabad, Iran

Abstract

This chapter describes a parametric iteration method for solving nonlinear Fredholm integro-differential equations (NFIDEs). Some examples illustrate the solution process. Comparisons are made between the Parametric Iteration Model (PIM) and the exact solution. The results show the simplicity and efficiency of PIM. Also, the convergence of this method is studied in this work.

Keywords: Parametric Iteration Method (PIM), nonlinear Fredholm integro-differential equations

1. Introduction

It is well known that many events in scientific fields deal with integro-differential equations. The nonlinear Fredholm integro-differential equations play a significant role in many physics, chemistry, biology, electrostatics, theory of elasticity, potential theory, astronomy, economics, and fluid dynamics (Mikhlin, 1964; Gavalas, 1968; Bird et al., 1962).

* Corresponding Author's Email: tamamgar_m@yahoo.com

In: Complexity and Fractality in Human Change and Beyond
Editor: Nicoletta Sala
ISBN: 979-8-89530-575-1

Recently, the analytical approximate and numerical methods have been proposed for the solution of nonlinear integro-differential equations (El-Sayed & Abdel-Aziz, 2003; Dehghan & Salehi, 2012; Akrami Arani & Sabzevari, 2023).

The PIM has been applied successfully to a class of integral equations (Maleknejad & Tamamgar, 2014; Tamamgar, 2014; Tamamgar, 2022). We consider nonlinear Fredholm integro-differential equations as the following:

$$u^{(m)}(x) = f(x) + \lambda \int_a^b k(x,t)F(u(t))dt \quad , \quad a \le x \le b \tag{1}$$

and

$$u^{(m)}(x) = f(x) + \lambda \int_a^b k(x,t)F(u,u')dt \quad , \quad a \le x \le b \tag{2}$$

Moreover, the initial value for both of the two equations is as follows:

$$u^{(i)}(x_0) = y_i, \quad (i = 0,1,\dots,m-1). \tag{3}$$

In this work, the numerical solution of (1) and (2) is possible by PIM when f, k, u are continuous and F is a continuous operator. The parametric iteration method provides a solution for NFIDEs as a sequence of iterations. In this chapter, some examples are given, and we solve them using the parametric iteration method. The results are compared with the exact solution. In all these cases, the present technique worked excellently, as will be shown in this study.

2. The Basic Idea of the Parametric Iteration Method

In this section, we describe PIM for solving nonlinear Fredholm integro-differential equations. Then, the convergence is discussed.

2.1. Parametric Iteration Method

The PIM provides the solution for Eq. (1, 2) as a sequence of approximations. This method gives rapidly convergent successive approximations of the exact solution if such a solution exists. Otherwise, approximations can be used for numerical purposes. We assume L and N are the linear and nonlinear operators on $C^m[a,b]$. To explain the basic idea of PIM, we consider Eq. (1, 2) as below:

$$L(u) + N(u) = f(x) \tag{4}$$

where L with the property $L(g) \equiv 0$ when $g \equiv 0$ denotes the so-called auxiliary linear operator concerning u, N is a nonlinear continuous operator for u and $f(x)$ is the known continuous function, and $u \in C^m[a,b]$.

The fundamental essence of this method is to construct a family of iterative processes for Eq. (1, 2) as:

$$L[u_{k+1}(x) - u_k(x)] = hH(x)A[u_k(x)], \tag{5}$$

with the initial conditions

$$u^{(i)}(x_0) = y_i, \quad (i = 0,1,\ldots,m-1)$$

where,

$$A[u_k(x)] = L[(u_k(x)] + N[u_k(x)] - f(x) = u_k^{(m)} - \lambda \int_a^b k(x,t)u_k^n\, dt - f(x),$$

or

$$A[u_k(x)] = L[(u_k(x)] + N[u_k(x)] - f(x) = u_k^{(m)} - \lambda \int_a^b F(x,t,u_k,u_k')dt - f(x), \tag{6}$$

and $u_0(x)$ is the initial guess that can be chosen arbitrarily but the suitable selection is positively affected by the rate of convergence, or it can also be solved from its corresponding linear homogeneous equation $L[u_0(x)] = 0$ or linear non-homogeneous equation $L[u_0(x)] = f(x)$. The parameter $h \neq 0$ and function $H(x) \neq 0$ denote the so-called auxiliary parameter and auxiliary function. The selection of $h, H(x)$ was described by Liao (2003). Furthermore, we have the freedom to select the auxiliary linear operator L, the auxiliary parameter h, the auxiliary function $H(x)$, and the initial approximation $u_0(x)$. Therefore, if the successive approximations

$u_k(x)$, $k \geq 0$where are obtained by PIM in terms of the auxiliary parameter h then the exact solution may be given by $u(x) = \lim_{k\to\infty} u_k(x)$.

According to Liao (2003), let $V = \{ u: u \in C^m[a,b] \}$be the solution space and denote $\{e_j(x): e_j(x) \in V, j = 0,1,\ldots\}$ the set of base functions. Hence, we can represent the solution in the series $u(x) = \sum_{j=0}^{\infty} \alpha_j e_j(x)$, where $\alpha_j$is a coefficient belonging to real numbers. As long as the set of base functions is determined, the auxiliary linear operator L, the initial approximation $u_0(x)$ and the auxiliary function H(x) must be chosen so that all solutions of the corresponding PIM equations (4) exist, and this set of base functions can express it. Now, to avoid expensive computational works for solving (1, 2) via PIM, it is straightforward to use the following set of base functions:

$$\{(x-a)^j | j = 0,1,\ldots\} \tag{7}$$

That $u(x) = \sum_{j=0}^{\infty} \alpha_j (x-a)^j$ (8)

where $\alpha_j \in R$ are unknown coefficients to be determined and$a \in R$. Now, we set the auxiliary operator L as the following:

$$L[u(x)] = u^{(m)}(x) \tag{9}$$

The initial guess is to form a combination of m-term of (7) i.e.,

$$u_0(x) = \alpha_0 + \alpha_1(x-a) + \alpha_2(x-a)^2 + \cdots + \alpha_j(x-a)^j. \tag{10}$$

According to (9), the initial conditions (3), and with due attention to $L[u_0(x)] = f(x)$, the coefficients $\alpha_0, \alpha_1, \alpha_2, \ldots, \alpha_m$ will be determined. Also, we set $H(x) = 1$, the selecting of H(x) is arbitrary. However, the suitable selection depends on the base functions for the solution (Liao, 2003), and we use the PIM processes to compute the approximation solutions of (1, 2).

2.2. The Valid Region of h

Assume we obtain a family of solution series in terms of the auxiliary parameter h using the PIM. We consider this solution as a function in terms of *h, x*, then we derive (once or more) this function concerning x in $x = \beta$ that $\beta \in [a,b]$, i.e., we let $U = G(x,h)$ be the solution of (1) or (2), then we set

$$\Omega = \frac{\partial^i G}{\partial x^i}|_{x=\beta}, \quad \beta \in [a,b], \quad (i = 1,2,..)$$

therefore Ω will be in terms of h; now we plot Ω curve, according to these h curves, it is easy to discover the valid region of h, which corresponds to the line segments nearly parallel to the horizontal axis. This region is called the valid region of h, which we note with R_h. We ensure the solution series converges for any $h \in R_h$.

2.3. Analysis of Convergence of the Parametric Iteration Formula

In this section, we examine the local convergence of approximate solutions offered by PIM for solving (1). The convergence of the approximate solution for (2) is similar to (1).

Initially, we let $u^{(i)}(0) = 0, \quad (i = 0,1,\dots,m-1)$ and set $L[u(x)] = u^{(m)}(x)$ therefore, we have from (5) the following parametric iteration formula:

$$u_{k+1}(x) = (1+h)u_k(x) - \frac{h}{(m-1)!}\int_0^x (x-t)^{m-1}\left[g(t) + \lambda\int_a^b k(t,s)F\big(u_k(s)\big)ds\right]dt \quad (11)$$

The iterative formula (11) expressed by sequence makes a recurrence sequence$\{\boldsymbol{u_k(x)}\}$. The limit of the sequence will be the solution of Eq. (1) if the sequence is convergent. To prove the sequence $\{\boldsymbol{u_k(x)}\}$ is convergent, we construct a series

$$u_0(x) + [u_1(x) - u_0(x)] + \dots + [u_k(x) - u_{k-1}(x)] + \dots \quad (12)$$

Noticing that

$s_{k+1} = u_0(x) + [u_1(x) - u_0(x)] + \ldots + [u_k(x) - u_{k-1}(x)] = u_k(x).$ (13)

The sequence $\{\boldsymbol{u_k}(\boldsymbol{x})\}$ will be convergent if the series is convergent.

Theorem: If $F(u(t))$ is Lipschitz-continuous in $[a, b]$ and $g(x) \in C[a, b]$ and

$|\lambda| \leq \frac{1}{MN}$ if *M* and *N* are positive real numbers, then the series of (12) is convergent, i.e., the sequence $\{u_k(x)\}$ is convergent for$x \in [a, b]$.

Proof: According to (11), note that

$$|u_1(x) - u_0(x)| = \left|h\left[u_0(x) - \frac{1}{(m-1)!}\int_0^x (x-t)^{m-1}\left[g(t) + \lambda \int_a^b k(t,s)F(u_0(s))ds\right]dt\right]\right| \leq \left|h\left[u_0(x) - \frac{1}{(m-1)!}\int_0^x (x-t)^{m-1}\left[g(t) + \frac{1}{MN}G(t)\right]dt\right]\right| \leq \left|h\left[u_0(x) - \frac{1}{(m-1)!}Y(x)\right]\right| \leq |h|[L_1 + L_2] = |h|r \quad (14)$$

where

$$L_1 = \max_{a \leq t \leq x \leq b} |u_0(t)|, L_2 = \max_{a \leq t \leq x \leq b} |(Y(x)|, L_1 + L_2 = r \quad (15)$$

From (11) and (14), and the assumption that $|F(u_{k+1}) - F(u_k)| \leq N|u_{k+1} - u_k|$ where *N* denotes the Lipschitz constant of $F(u(t))$, it follows that

$$|u_2 - u_1| = \left|\left[(1+h)(u_1 - u_0) - \frac{h}{(m-1)!}\int_0^x (x-t)^{m-1}\left[\lambda \int_a^b k(t,s)(F(u_1) - F(u_0))ds\right]dt\right]\right| \leq \left|\, |1+h||u_1 - u_0| + \frac{|h|}{(m-1)!}\int_0^x (x-t)^{(m-1)}\left[\lambda N \int_a^b k(t,s)|u_1 - u_0|ds\right]dt\right| \leq |h|r\left[|1+h| + \frac{L_2|h|}{(m-1)!}\right] = |h|r\sum_{n=0}^{1}\binom{1}{n}|1+h|^{1-n}\left(\frac{L_2|h|}{(m-1)!}\right)^n \quad (16)$$

$$|u_3 - u_2| = \left|\left[(1+h)(u_2 - u_1) - \frac{h}{(m-1)!}\int_0^x (x-t)^{m-1}\left[\lambda \int_a^b k(t,s)(F(u_2) - F(u_1))ds\right]dt\right]\right|$$

$$\leq \left| |1 + h||u_2 - u_1| + \frac{|h|}{(m-1)!} \int_0^x (x-t)^{(m-1)} \left[\lambda N \int_a^b k(t,s)|u_2 - u_1| ds \right] dt \right|$$

$$\leq |h|r \left[|1 + h|^2 + \frac{2L_2|h||1+h|}{(m-1)!} + \frac{{L_2}^2|h|^2}{((m-1)!)^2} \right] = |h|r \sum_{n=0}^{2} \binom{2}{n} |1 + h|^{1-n} \left(\frac{L_2|h|}{(m-1)!} \right)^n \quad (17)$$

$\vdots$

$$|u_{k+1} - u_k| \leq (|h|r) \sum_{n=0}^{k} \binom{k}{n} |1 + h|^{k-n} \left(\frac{L_2|h|}{(m-1)!} \right)^n \quad (18)$$

Given (18), the convergence of the series (12) can be concluded for the solution domain $x > 0$ and $|1 + h| < 1$ with the help of some mathematical software. Therefore, the series of (12) is absolute convergence, i.e., the sequence $\{u_k(x)\}$ is convergent for $x \in [a, b]$.

3. Illustrative Examples

Now, we use PIM to solve two examples of type (1) and type (2) and compare the results obtained with ADM (Liao, 2003) to show the efficiency of PIM.

Example 1. The first example is a nonlinear Fredholm integro-differential of the second kind as follows:

$$u'(x) = 1 - \frac{x}{3} + \int_0^1 x u^2(t) dt \quad (19)$$

$$u(0) = 0 \quad (20)$$

According to PIM proceeding, we define:

$$L[u(x)] = u'(x) \quad (21)$$

and

$$N[u(x)] = \int_0^1 x u^2(t) dt, \quad f(x) = 1 - \frac{x}{3} \quad (22)$$

Then we have from (8) $u_0(x) = \alpha_0 + \alpha_1 x$. Using initial conditions and $L[u_0(x)] = f(x)$ gives us that$\alpha_0 = 0, \alpha_1 = -1$ i.e., $u_0(x) = -x$. Now, If we set $H(x) = 1$ then we obtain from (5), (6),

$$u'_{k+1}(x) - u'_k(x) = h(u'_k(x) - \int_0^1 x u_k^2(t)dt + 1 - \frac{x}{3}) \quad (23)$$

Thus, we integrate from both sides of (24), and then the PIM equations are as follows:

$$u_{k+1}(x) = (1 + h)u_k(x) - h\int_0^x \left[1 - \frac{t}{3} + \int_0^1 t u_k^2(s)ds\right)]dt, (k = 0,1, \dots) \quad (24)$$

The obtained result for the third iteration is the exact solution, i.e., $u(x) = x$ for$h = -1$; also the valid region of h, i.e., R_h was presented in Figure 1.

Example 2. We consider the nonlinear Fredholm integro-differential of the second kind as the following:

$$u''(x) = 6x + \int_{-1}^1 xt\, u(t)u'(t)dt \quad (25)$$

$$u(1) = u(-1) = 0 \quad (26)$$

The exact solution is $u(x) = x^3 - x$. According to the PIM procedure we set,

$$L[u(x)] = u''(x) \quad (27)$$

and

$$N[u(x)] = \int_{-1}^1 xtu(t)u'(t)dt, \quad f(x) = 6x \quad (28)$$

Then we have from (8) $u_0(x) = \alpha_0 + \alpha_1 x + \alpha_2 x^2 + \alpha_3 x^3$, (27), (28) gives us that

$$\alpha_0 = 0, \alpha_1 = -1, \alpha_2 = 0, \alpha_3 = 1$$

i.e., $u_0(x) = x^3 - x$. Now, similar to example 1, the iteration scheme is as follows:

$$u_{k+1}(x) = (1 + h)u_k(x) - h\int_0^x (x - t)[6t + \int_{-1}^{1} ts\, u_k(s)\, u_k'(s)ds)]dt, (k = 0,1, \dots) \quad (29)$$

The results of Example 2 are available in Table 1. R_h for Eq. (25) is presented in Figure 2.

Table 1. The results of Example 2 for $u_8(x)$ with $h = -0.0001$

x	Exact solution	PIM
0.0000	0.0000	0.0000
0.1	-0.099	-0.09892002800
0.2	-0.192	-0.1918400560
0.3	-0.273	-0.2727600840
0.4	-0.336	-0.3356801120
0.5	-0.375	-0.3746001400
0.6	-0.384	-0.3835201680
0.7	-0.357	-0.3564401960
0.8	-0.288	-0.2873602240
0.9	-0.171	-0.1702802520
1.0000	0.0000	0.0007997200

Table 2. The results of Example 3 for $u_3(x)$ with $h = -1$

x	Exact solution	PIM
0.0000	0.0000	0.0000
0.1	0.1	0.09888868261
0.2	0.2	0.1978084873
0.3	0.3	0.2967905361
0.4	0.4	0.3958659512
0.5	0.5	0.4950658546
0.6	0.6	0.5944213684
0.7	0.7	0.6939636146
0.8	0.8	0.7937237154
0.9	0.9	0.8937327927
1.0000	1.0000	0.9940219688

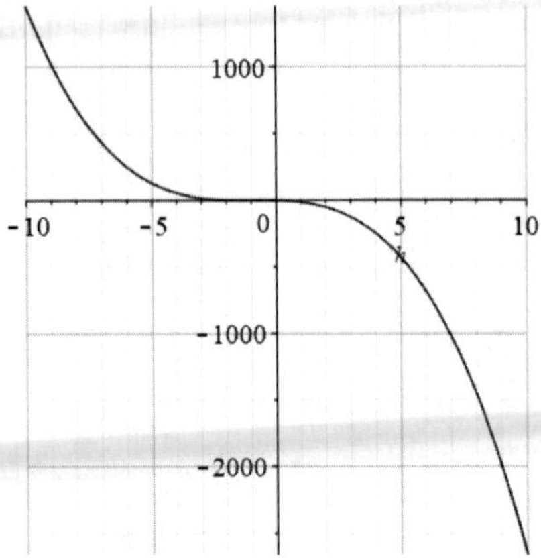

Figure 1. The valid region of *h*, for Example 2 shows $-1.8 \leq h \leq 0.8$.

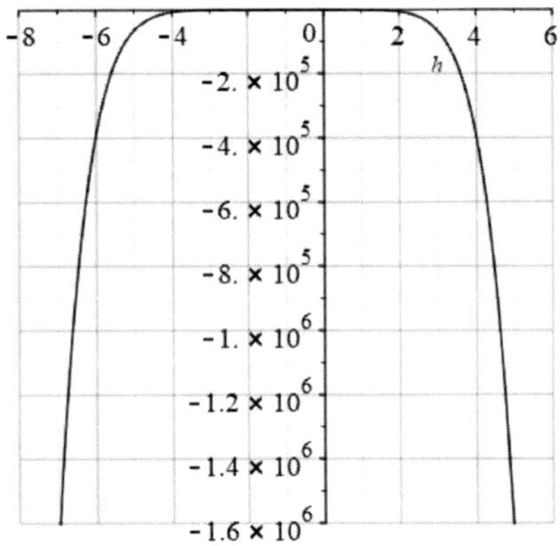

Figure 2. The valid region of *h*; for Example 2 shows $-3.5 \leq h \leq 1.5$.

Example 3. We consider the nonlinear Fredholm integro-differential of the second kind as the following:

$$u'(x) = \frac{5}{4} - \frac{x^2}{3} + \int_0^1 (x^2 - t)u^2(t)dt \tag{30}$$

That $u(0) = 0$, (31)

The exact solution is $u(x) = x$. We set:

$$L[u(x)] = u'(x) \tag{32}$$

and

$$N[u(x)] = \int_0^1 (x^2 - t)u^2(t)dt, \quad f(x) = \frac{5}{4} - \frac{x^2}{3} \tag{33}$$

Now, we consider the initial solution as $u_0(x) = \alpha_0 + \alpha_1 x + \alpha_2 x^2 + \alpha_3 x^3$, (31), (32) gives us that

$$\alpha_0 = 0, \alpha_1 = \frac{5}{4}, \alpha_2 = 0, \alpha_3 = -\frac{1}{9}$$

i.e., $u_0(x) = \frac{5}{4}x - \frac{1}{9}x^3$. Now, similar to examples 1 and 2; the iteration scheme is as the following:

$$u_{k+1}(x) = (1 + h)u_k(x) - h\int_0^x [\frac{5}{4} - \frac{t^2}{3} + \int_0^1 (t^2 - s)u_k^2(s)ds)]dt, \quad (k = 0,1,\dots) \tag{34}$$

The results of Example 3 are available in Table 2. $R_h$for Eq. (30) is presented in Figure 3.

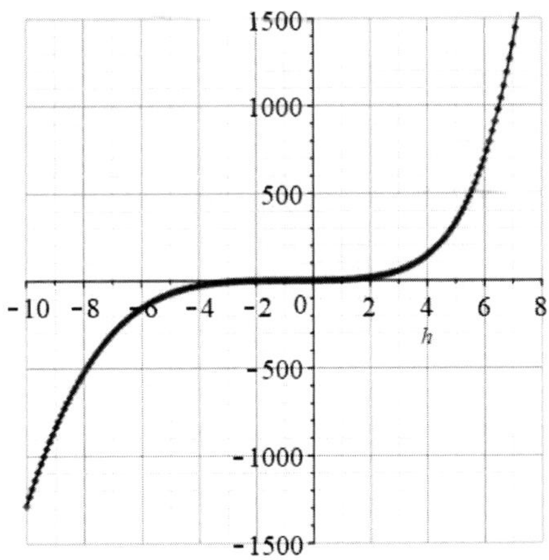

Figure 3. The valid region of h; for Example 2 shows $-3 \leq h \leq 1$.

4. Results and Figures

The results of examples 2, and 3 in two tables (1 and 2) and plot the h-curve to determine R_h (Figures 1, 2 and 3) are described in the previous section. All the computations were performed using Maple 2019.

Conclusion

In this chapter, we applied the PIM to solve the nonlinear Fredholm integro-differential equations. To illustrate this method, we solve three examples. Compared to the exact solution, PIM results indicate it is more straightforward in practice and more accurate for NFIDEs. In the first example, the exact solution was obtained.

Further, the convergence of PIM for solving NFIDEs in the valid region of h (R_h) was presented. Additionally, increasing the number of iterations through the PIM scheme appears to result in more accurate solutions.

References

Akrami Arani, M., & Sabzevari M. (2023). Numerical solution of system of nonlinear Fredholm integro-differential equations using CAS wavelets, *Journal of Mathematical Modeling*, 11 (3), 587–602.

Bird, R. B, Stewart, W. E, & Lighfoot, E. N. (1962). *Transport Phenomena*, Wiley, New York, London.

Dehghan, M., & Salehi, R. (2012). The numerical solution of the non-linear integro-differential equations based on the meshless method. *Journal of Computational and Applied Mathematics*, 236 (9), 2367–2377.

El-Sayed, S. M., & Abdel-Aziz, M. R. (2003). Comparison of Adomians decomposition method and wavelet-Galerkin method for solving integro-differential equations. *Appl. Math. Comput*. 136 (1), 151-159.

Gavalas, G. R. (1968). *Nonlinear Differential Equations of Chemically Reacting Systems*. Springer- Verlag, Berlin, Heidelberg, New York.

Maleknejad, K., & Tamamgar, M. (2014). A new reconstruction of variational iteration method and its application to nonlinear Volterra integro-differential equations. *Abstr. Appl. Anal.* (SI28) 1 - 6.

Mikhlin, S. G. (1964). Integral Equations and Their Applications to Certain Problems. *Mechanics, Mathematical Physics and Technology,* Second edition, Pergamon, Press, New York

Liao, Sh. J. (2003). *Beyond perturbation: Introduction to the Homotopy Analysis Method.* Boca Raton: Chapman & Hall/CRC Press.

Tamamgar, M. (2022). An effective method for solving the Nonlinear Fredholm Integral Equations, *Chaos and Complexity Letters*, 16, 85-94.

Tamamgar, M. (2014). The Numerical Solution of linear Fredholm equations via Parametric iteration Method. *Appl. and Comput. Math.*, 3, 4-10.

Chapter 4

Saturation Points in Some Collatz Sequences

Danilo Merlini[1,2]
and Nicoletta Sala[2,3,*]

[1]Research Center for Mathematics and Physics, CERFIM, Locarno, Switzerland
[2]Institute for Scientific and Interdisciplinary Studies, ISSI, Locarno, Switzerland
Institute in Arithmetic Physics, IRFIA, Locarno, Switzerland
[3]Institute for Complexity Studies, Rome, Italy

Abstract

We analyze the orbits in some $3 \cdot n + a$ Collatz sequences for a finite set of the integers n, i.e., for the cases a = 1, a = 7, and a = -1.
For a = -1 and a = 7, we observe a "saturation point" as already introduced and computed in previous work for the case a = 1.
For a = -1, the saturation point is $n_s = 3$, and for a = 7, it is given by $n_s = 27$. Then, for these two cases (a = -1 and a = 7), we check the truth of a conjecture (similar to that previously stated for the case a = 1 in a finite domain in the ranges of $n \in (1, 1000)$) that for k (the depth of the trees of the inverse orbits defined below), there are at least the first k integers on the trees.
Then, we continue the analysis of the inverse orbits for the main case, i.e., a = 1, looking at the following:

a) The total density of the odd numbers on the tree at depth k in connection with stochastic models.
b) The linear density on the chalice (tree) at a fixed depth k.
c) The density of an inverse orbit ("vertical"), i.e., the density of the odd to the even numbers in some special generic inverse orbits.

---

* Corresponding Author's Email: nicolettasalavb@gmail.com

In: Complexity and Fractality in Human Change and Beyond
Editor: Nicoletta Sala
ISBN: 979-8-89530-575-1

Finally, we propose a condition on the orbits for the (3·n + 1)/2 case such that the saturation point at n = 73 disappears.

Keywords: Collatz problem, orbits, stopping time, saturation points, some numerical experiments from

1. Introduction

The 3·n + 1 or (3·n + 1)/2 problem is characterized by having only a small cycle, likely involving the convergence of the orbits of all integers n. It is given by 4→2→1→4 or 2→1→2.

For the 3·n + 1 formulation, we recall that the sequence is defined by: if n is an odd integer, compute 3·n + 1; if n is an even integer, compute n/2, and so on.

There is no complete proof that all integers eventually fall into cycle 1, 4, and 2. It remains possible that an infinite set of integers does not enter this cycle; some trajectories diverge to infinity, or some integers belong to a much larger cycle that includes many odd integers. See the extensive work of Lagarias for important explanations of many types of models related to the 3n + 1 (Lagarias, 2003; Lagarias, 2006).

A point of interest is that all similar problems, i.e., 3·n + a, where a is odd and n > 0, possess the elementary cycle (multiple of the previous 3·n + 1 problem), i.e., a→4·a→2·a→a, which corresponds to the occurrences of "all" multiples of 3, (when a = 3), "all" multiples of 5 (when a = 5), "all" multiples of 7, and so on, in addition to other potentially more significant cycles (Section 3 illustrates some trees).

Experiments can still be interesting for finite sets of integers that are not necessarily large, as illustrated by a set that contains one thousand integers (1000). From a list of the lengths of the orbits calculated up to n = 1000, for the usual case, a = 1, a previous analysis of the orbits has revealed the emergence of a "saturation point" located in the 3·n + 1 formulation at n = 118, and for the (3·n + 1)/2 at n = 73 (Merlini & Sala, 2022). The saturation point is defined to be such that the length l(n) of the orbit of an integer n reaching 1 (in the case a = 1) or achieving one of the numbers belonging to one of the cycles of the model for a (3 cycles for a = -1 and 2 cycle for a = 7, the two cases analyzed here) is equal to the number n, thus $l(n_s) = n_s$ and $l(n) < n$ for all $n > n_s$, with n in the finite domain we consider.

Equivalently, the tree of the inverse orbits of depth k is expected to contain all numbers from 1 to k, giving rise to the abovementioned conjecture.

We presented the list and calculated the lengths of the orbits up to n = 1000, from which we directly checked the correctness of the conjecture for the finite domain above in case a = 1. We observed that as the intervals of n grow, i.e., given by [1, 250], [250, 500], [500, 750], and [750, 1000], the ratio between the length of the most extended orbits over n, i.e., the ratio l(n)/n, decreases as a function of the center of the intervals, a symptom that the conjecture may continue to be true as n increases (Merlini & Sala, 2022).

We then reconsider the case a = 1, where we presented the original chalice (the tree of the inverse orbits in the 3·n + 1 formulation) of depth k = 15 (Merlini & Sala, 1999; Merlini & Sala, 2022).

We first examine a standard bound for the density of even and odd numbers in the sequences. Then, we conduct a numerical experiment to calculate the density of odd and even numbers in specific orbits.

In these cases, the density (the ratio between the odd and the even) approaches a value near 1 in the (3·n + 1)/2) formulation, from below and from above, and the value near ½ in the 3·n + 1 formulation. This may be of interest in discussing the possible inexistence of isolated cycles in the 3·n + 1 Collatz sequences.

2. Equation for Cycles Containing Very Few Odd Numbers (1, 2, 3)

We start with the Equation for tiny cycles in the 3·n + a problem, where a, is an odd constant.

2.1. Cycles Containing One Odd (Let d_1 Be a Positive or a Negative Integer)

We have:

$$\frac{(3 \cdot d_1 + a)}{2^{\alpha_1}} = d_1$$

with the solution

$$d_1 \cdot (2^{\alpha_1} - 3) = a \quad (1)$$

If $\alpha_1 = 2, \rightarrow d_1 = a$.

For $a = 1$, $d_1 = 1$, cycle $1 \rightarrow 4 \rightarrow 2 \rightarrow 1$ $(3 \cdot n + 1)$

For $a = 5$, $d_1 = 5$ cycle $5 \rightarrow 20 \rightarrow 10 \rightarrow 5$ $(3 \cdot n + 5)$

If $\alpha_1 = 3, \rightarrow 5 \cdot d_1 = a$.

$a = 5$, $d_1 = 1$, cycle $C_{1,5}$ is $1 \rightarrow 8 \rightarrow 4 \rightarrow 2 \rightarrow 1$ $(3 \cdot n + 5)$

For

$a = 7$, $d_1 = 7$, cycle $C_{1,7}$ is $7 \rightarrow 28 \rightarrow 14 \rightarrow 7$ $(3 \cdot n + 7)$. (2)

In the cycle $C_{1,a}$ if a is negative, for example, $a = -1$, we have:

$\alpha_1 = 1$ $a = -1$, $d_1 = 1$, cycle $C_{1,-1}$ $1 \rightarrow 2 \rightarrow 1$ $(3 \cdot n - 1)$.

2.2. A Cycle Containing Two Odd Numbers, d_1 and $d_2 = (3 \cdot d_1 + a)/2^{\alpha 1}$

We have:

$$\frac{3 \cdot \left[\frac{(3 \cdot d_1 + a)}{2^{\alpha_1}}\right] + a}{2^{\alpha_2}} = d_1$$

and

$$d_1 \cdot (2^{\alpha_1 + \alpha_2} - 3^2) = a \cdot (3 + 2^{\alpha_1}) \quad (3)$$

For $\alpha_1 + \alpha_2 = 4$, $7 \cdot d_1 = a \cdot 5$, $\rightarrow d_1 = 5$, $a = 7$. ($\alpha_1 = 1$ and $\alpha_2 = 3$)

Cycle $C_{2,7}$:

$5 \rightarrow 22 \rightarrow 11 \rightarrow 40 \rightarrow 20 \rightarrow 10 \rightarrow 5$ $(3 \cdot n + 7)$ (4)

If $\alpha_1 = 1$ and $\alpha_2 = 2$, we have $\alpha_1 + \alpha_2 = 3$, $- d_1 = a\cdot(3 + 2^{\alpha 1})$, for $a = -1$

$-d_1 = -(3 + 2^{\alpha 1})$, $\alpha_1 = 1$, $d_1 = 5$, cycle 5→14→7→20→10→5

If $\alpha_1 + \alpha_2 = 3$, for $\alpha_1 = 2$, $\alpha_2 = 1$, the cycle is cycle 7→20→10→5→14→7.

2.3. Cycles Containing Three Odd Numbers, d_1, $d_2 = \frac{(3\cdot d_1 + a)}{2^{\alpha_1}}$ $d_3 = \frac{(3\cdot d_2 + a)}{2^{\alpha_2}}$ and $\frac{(3\cdot d_3 + a)}{2^{\alpha_3}} = d_4 = d_1$

That is:

$$d_1 \cdot (2^{\alpha_1 + \alpha_2 + \alpha_3} - 3^3) = a \cdot (9 + 3 \cdot 2^{\alpha_1} + 2^{\alpha_1 + \alpha_2}) \quad (5)$$

For $\alpha_1 + \alpha_2 + \alpha_3 = 5$

$$d_1 \cdot 5 = a \cdot (9 + 3 \cdot 2^{\alpha_1} + 2^{\alpha_1 + \alpha_2})$$

If $\alpha_1 = \alpha_2 = 1$, $\alpha_3 = 3 \rightarrow d_1 = 19$ with $a = 5$.

Cycle $C_{3,5}$ with three odd numbers:

19→62→31→98→49→152→76→38→19 $(3\cdot n + 5)$ (6)

If $\alpha_1 = 1$, $\alpha_2 = 2$ and $\alpha_3 = 2$, $d_1 \cdot 5 = a \cdot (9 + 3 \cdot 2^{\alpha_1} + 2^{\alpha_1 + \alpha_2}) = a \cdot$
2

with the solution $d_1 = 23$ and $a = 5$.

Cycle with three odd numbers:

23→74→37→116→58→29→92→46→23 (7)

In Figure 1, we present the trees of case $3\cdot n + 7$, i.e., the trees starting from the cycles $C_{1,7}$ (containing the multiple of 7) and the other cycle $C_{2,7}$ 5→22→11→40→20→10→5 containing two odd numbers; this illustrates the connection with the case of primary interest here, i.e., the cycle 1→4→2→1 of the $3 \cdot n + 1$.

2.4. An Example of a Cycle Containing 7 Odd Numbers

It is the case for a = -1, i.e., where we have three cycles, i.e., 1→2→1, ($C_{1,-1}$), the cycle 5→14→7→ 20→10 →5, ($C_{2,-1}$) and the cycle 17→50→25→74→37→110→55→164→82→41→122→61→ 182→91→ 272→136→68 →34→17, of length l = 18 ($C_{7,-1}$). (The second index refers to a, i.e., here to a = -1).

Remarks

1) Of course, for d_1 = a, i.e., $3 \cdot n + d_1$, we have the cycles d_1, $4 \cdot d_1$, $2 \cdot d_1$, d_1, for every odd d_1.

All the corresponding trees are the same as that for $d_1 = 1$, i.e., $3 \cdot n + 1$ where each number in the tree of the $3 \cdot n + 1$ case is multiplied by d_1, that is the tree of "all" multiples of 3, that of "all" multiples of 5, that of "all" multiples of 7 and so on, while for d_1 = - a, we have the cycles $d_1 \to 2 \cdot d_1 \to d_1$ for every odd d_1.

2) It has been observed that for negative integers, there is not "only" one cycle but three cycles, i.e., -1→-2→-1, -5→-14→-7→-20→-10→-5 and the cycle containing seven odd numbers presented above: -17→-50→-25→-74→-37→-110→ -55→ -164→ -82→ -41→ -122→ -61→ -182→ -91→ -272→ -136→ -68→-34→-17 (Lagarias, 1985; Bishoff, 2024).

We remark here that these are the cycles of the Collatz sequences for a = -1 presented in the above lines, i.e., a change of the signum of all numbers of the sequences obtained with the change a = 1→ a = -1. Thus, it is no mystery that for negative numbers, the $3 \cdot n + 1$ has three cycles, as has been observed in many other cases, since they arise for a = -1, but for n > 0. For extensive analysis and interesting computations, see Lagarias (1985).

3. Saturation Point and the Conjecture in the Three Cases: a = 7, a = -1, and a = 3

We noticed that the list of the length of an orbit falling in one of the cycles is defined to be zero if the number belongs to the cycle and for other numbers of the orbit's end reaching one of the numbers of the cycle. (As an example for a = 7, in the orbit of n = 27, i.e., 27 → 88 → 44 → 22 → 11 → 40 → 20 → 10 → 5 → 22 → 11 → 40 → 20 → 10 → 5 the length is only l(27) = 3 since 22 belongs to the cycle. One can, of course, define the length in another way, for example, by the arrival of the orbit at the smallest integer in the cycle (in the above example, n = 5), but this will be immaterial for the "value" of a saturation point to be found below.

3.1. The Case 3·n + 7

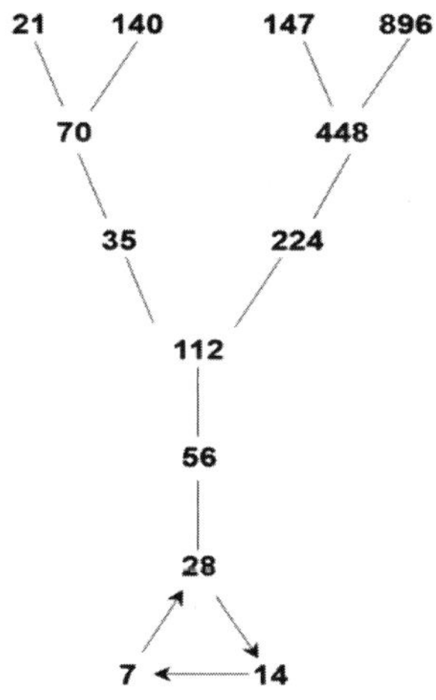

Figure 1. The tree $C_{1,7}$ in the 3·n + 7.

Here, the orbits split into two sets: "all" multiples of 7, which fall into the cycle 7 → 28 → 14 → 7 ($C_{1,7}$), and "all" the other integers, which fall into the cycle $C_{2,7}$, i.e., 5 → 22 → 11 → 40 → 20 → 10 → 5. We notice here that in Wegner (2021), the computations have reached n = 107 with starting argument x_o, and for the 3·n + 7 case, only one additional cycle to $C_{1,7}$, i.e., $C_{2,7}$, has been found. In this case, the plot of the orbits' length below allows us to localize a saturation point at n_s = 27. The point plot of the length of the orbits up to n = 60 is shown in Figure 1. The length of any number on the cycle is equal to zero in the cycles $C_{1,7}$ and $C_{2,7}$ (Figure 2). Figure 3 illustrates the point plot of l(n) for the case (3·n + 7), and Figure 4 shows the plot of N(k).

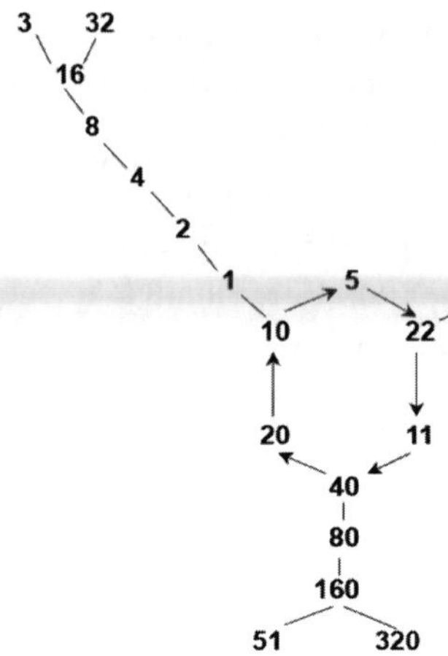

Figure 2. The tree $C_{2,7}$ in the 3·n + 7.

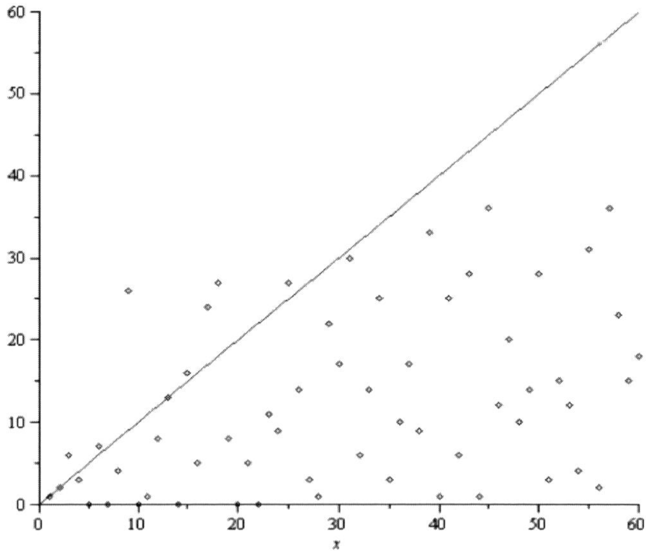

Figure 3. Point plot of l(n) for the case (3·n + 7). Above n = 27, "all" points from n = 28 are below the line y = n: the saturation point is located at n_s = 27 as specified below (the range is here (1..60)).

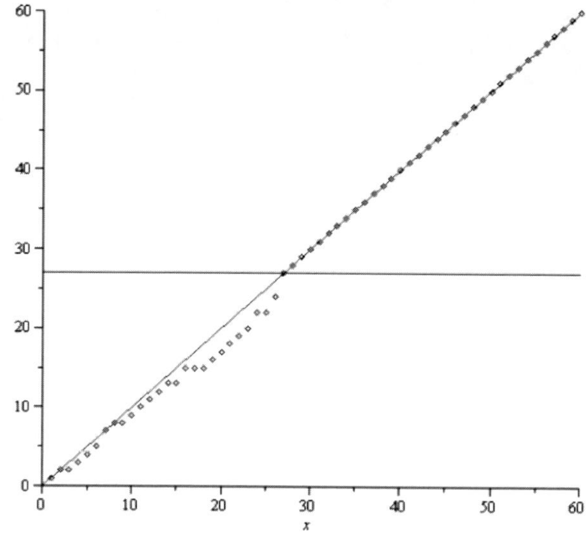

Figure 4. The plot of N(k), i.e., the number of the integers not exceeding k appearing in the trees of the inverse orbit of (3·n + 7), as a function of the depth k (for both trees) in the range k = [0..60].

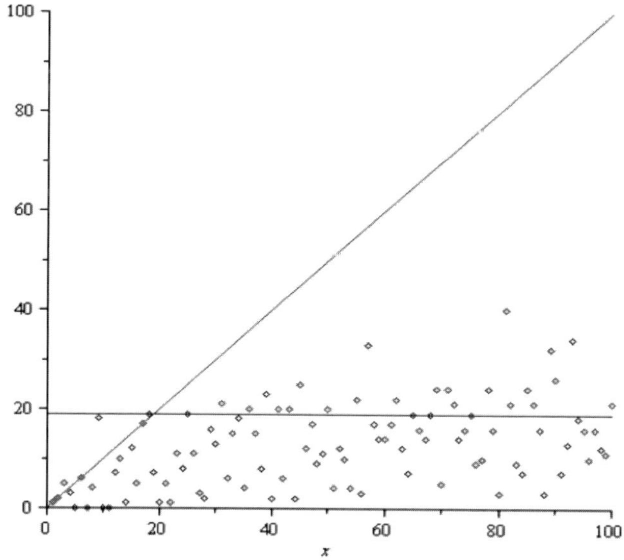

Figure 5. The plot of l(n) in the range n = [0..100]. Saturation point at n = 18 ((3·n + 7)/2).

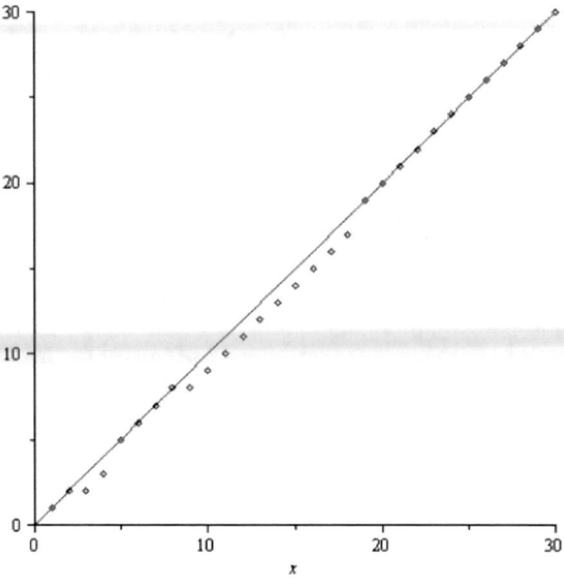

Figure 6. The plot of N(k) in the range [0..30] "all" points above n = 18 are on the plot of y = n. (N(19) = 19 even if here l(17) = 17).

Here the conjecture is:

For the two sets of the inverse orbits (trees of depth k), there are the first k numbers (1,2,...k) appearing on the two trees of the inverse orbits for k> k_s where k_s = 27 is the saturation point. Notice that for the first tree, i.e., with the cycle (7, 28, 14, 7), for the "all" multiple of 7, up to k = 60, we only have the short set given by (7, 14, 21, 28, 35, 42, 49).

We may also consider the second formulation, i.e., n→n/2 if n is even, and n→(3·n + 7)/2 if n is odd. Figure 5 presents the plot of l(n) and N(k) in this formulation. Figure 6 displays the plot of N(k) within the range [0..30].

3.2. The Case 3·n-1

In this case, we have three cycles, and the saturation point is given by n_s = 3. In fact, with the definition above, we have not found any n for which l(n) > n; in Figure 7, the plot of l(n) for this case in the formulation (3·n-1)/2 analogous to that of Figure 4, but all points are below the line given by y = n. Figure 8 shows the plot of N(k) as a function of the depth k.

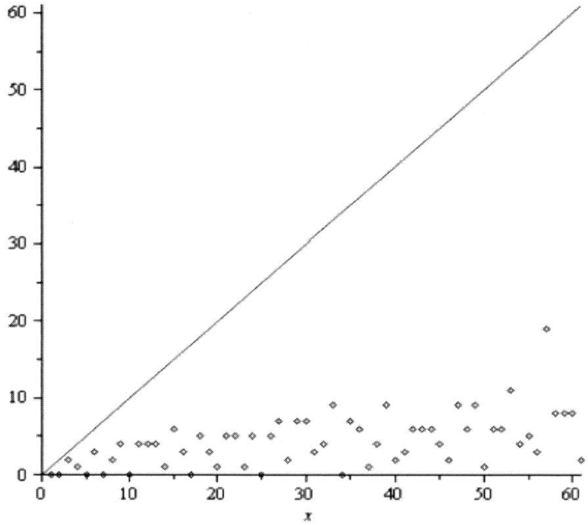

Figure 7. Point plot of l(n) for the case (3n-1)/2 in the range [0..61]. Here, all points (n, l(n)) are below the line y = n, and there is no saturation point (notice that the range is still [1..60]) (orbit of 3: 3→4→2, l(3) = 2).

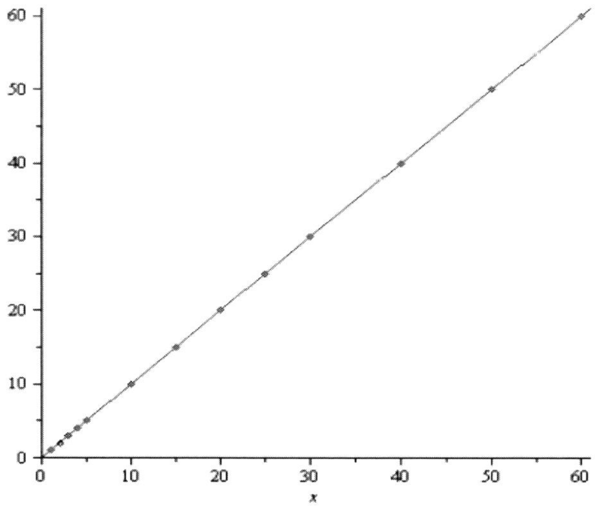

Figure 8. Point plot of N(k) as a function of the depth k. All points are on the line y = n.

3.3. The Case 3·n + 3

In this case, we have the cycle 3·(4, 2, 1) = (12, 6, 3) and the cycle 3·(2, 1) = (6, 3) in the (3·n + 3)/2 formulation.

If the Collatz conjecture is valid for the 3·n + 1, then the tree of 3·n + 3 is identical to that of all numbers multiplied by 3, meaning all multiples of three are included. As an example, the orbit of 3 in the 3·n + 1 is given by (3, 10, 5, 16, 8, 4, 2, 1), and in the 3·n + 3, the orbit of 9 is given by (9, 30, 15, 48, 24, 12, 6, 3) = 3·(3, 10, 5, 16, 8, 4, 2, 1). For the 3·n + 3, we do not know of any other existing cycle. If no cycles exist besides (12, 6, 3), then all results for 3·n + 1 apply to 3·n + 3. We have the conjecture:

The tree of the inverse orbit of the 3·n + 3 is "three-time" the tree of the 3·n + 1, i.e., each inverse orbit of the 3·n + 3 is the orbit of the 3·n + 1, each one multiplied by 3, as given above for the inverse orbit of n = 9. (This is true, of course, if we neglect the numbers that are not multiples of 3 but whose orbit falls into one that is a multiple of 3, as demonstrated below.) Figure 9 illustrates the plot of l(n) as a function of n in the formulation (3·n + 3)/2. Figure 10 displays the plot of N(k) as a function of the depth k and the constant f(k) = 71.

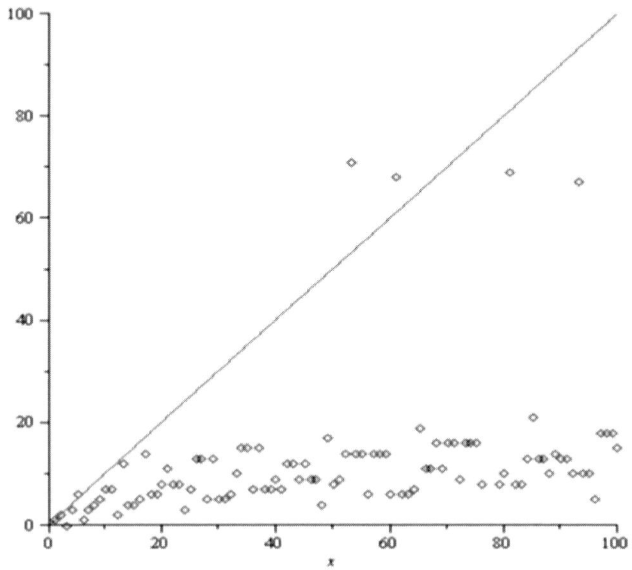

Figure 9. Plot of l(n) as a function of n in the formulation (3·n + 3)/2.

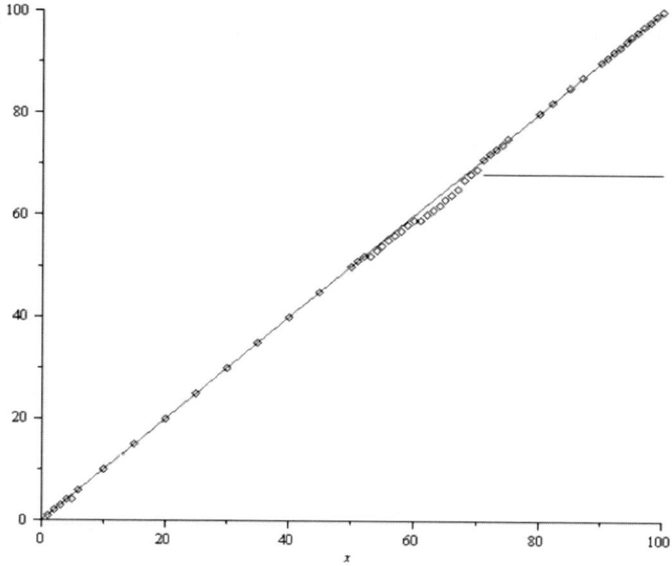

Figure 10. The plot of N(k) as a function of the depth k and the constant f(k) = 71. The saturation point is located at k = 71 (essentially the same as for the (3·n + 1)/2 cases where such a point has been localized at k = 73, i.e., l(73) = 73. See the above Remark).

Remark:

On the plot above the 4 points may be explained as follows: the four points are given by: (53,71) (2·27 = 54, in the (3n + 1)/2 (l(27) = 70)), (61, 68) (2·31 = 62, in the (3·n + 1)/2, (l(3) = 68)), (81,69) (3·27 = 81, (l(27 = 70), (2·41 = 82, l(41) = 69) in the (3·n + 1)/2), and l(93) = 67 (3·31 = 93 and l(31) = 67 in the (3·n + 1)/2 formulation). This explains the connection between 3·n + 1 and 3·n + 3 (the numbers 53, 61, 81, 93) are precise, i.e., near the multiples of the numbers (27, 31, 41), numbers that, in the (3·n + 1)/2 case, have the most extended orbits in the range n = (1..100).

Remark:

If a number n is not a multiple of 3, then n = 3·k + a where a = 1 or a = 2. Applying the rule once or twice results in a multiple of 3.

We have the following cases.

Case 1

$n = 3 \cdot k + 1$, k even, $3 \cdot k + 1$ odd $\rightarrow 3 \cdot (3 \cdot k + 1) + 3$ multiple of 3; k odd, $3 \cdot k + 1$ even, $(3 \cdot k + 1)/2^{\alpha}$ is odd for some α and $3 \cdot ((3k + 1)/2^{\alpha}) + 3$ multiple of 3.

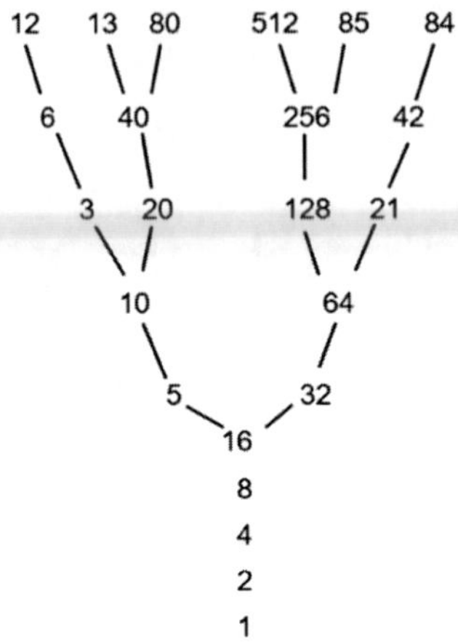

Figure 11. Tree of depth 9 in the $3 \cdot n + 1$.

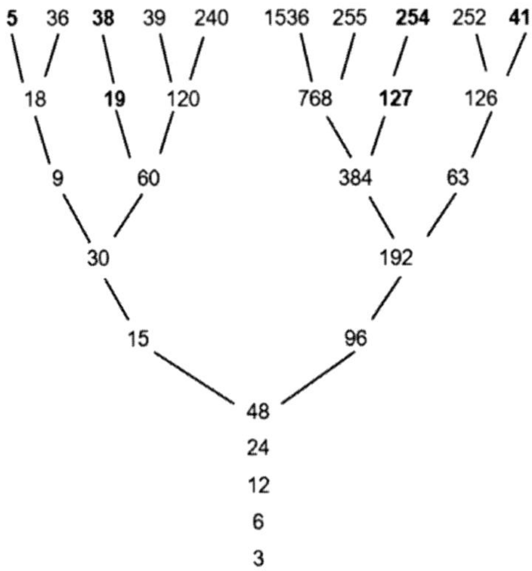

Figure 12. A complete tree of depth 9 in $3 \cdot n + 3$. The figure includes the six numbers, that are not multiples of 3, in bold.

Case 2

$n = 3k + 2$, k even, $3k + 2$ even → $(3k + 2)/2^{\beta}$ is odd for some β and $3\cdot((3k + 2)/2^{\beta}) + 3$ is a multiple of 3; k odd, $3k + 2$ odd→$3\cdot(3k + 2) + 3$ is a multiple of 3.

Therefore, if there is no divergent trajectory in the $3\cdot n + 1$ and no other cycle, then (1, 4, 2, 1) applies similarly to the $3\cdot n + 3$, where the corresponding numbers are 3 times those of the $3\cdot n + 1$, after "neglecting" those that are not multiples of 3, which fall into the orbit of a multiple of 3 as proven above. A small tree of depth nine is given below in the two cases (Figure 11 and Figure 12).

The following section analyzes some special orbits in the original $3n + 1$ or $(3n + 1)/2$ sequences.

4. Density Analysis in Some Orbits of the 3·n + 1 Sequence

4.1. Relation on the Length of the Orbits in the Two Formulations, i.e., for the 3·n + 1 and (3·n + 1)/2

If l(n) and l'(n) are the lengths of the orbits reaching 1 in the $3\cdot n + 1$ resp. in the $(3\cdot n + 1)/2$ formulation then:

$$l'(n) = l(n) - (n_o-1) \qquad (8)$$

where n_o is the number of the odd in both the formulations (for example, $l(3) = 7$, $n_o = 3$ and $l'(3) = 5$ (3→5→8→4→2→1) and $5 = 7-(3-1)$.

For the computation of both l and l,' we refer to the Good Calculator (2024).

4.2. Densities (Ratio Between the Odd and the Even Numbers)

4.2.1. Density of the Odd and Even Numbers in the Full Chalice

It is tempting to compute the density of odd and even numbers for large k in the model related to a Fibonacci sequence studied some time ago (Merlini & Sala, 1999; Merlini & Sala, 2022). In that model, it was found (in the $3\cdot n + 1$

formulation) that the total population on the chalice of depth k is given by $N_k = A.(c^{k+1}-1)/(c-1)$ where A ~ 0.654... and c = 1.2637... (notice that c is a solution of the equation c^2-c-1/3 = 0).

The quantity of the odd numbers in the "full chalice" (equal to the number of outcoming branches n_{k0} on the top of the "chalice," see the chalice presented in Figure 13 where k = 15, $n_{15,0}$ = 24 and N_{15} = 103), is given by $A \cdot c^k$.

Figure 13. Our original chalice of depth 15.

Since $n_{ko} + n_{ke} = N_k$ where n_{ke} is the number of even numbers, we have that, as $k \to \infty$, $\rho_{k,o} = n_{k,o}/N_k = (c-1)/c$ and $\rho_{k,e} = n_{k,e}/N_k = 1/c$ (where: $\rho_{k,o} + \rho_{k,e} = 1$).

Thus:

$\rho_{k,o} = 1 + (3/2)\cdot(1-(7/3)^{1/2}) \sim 0.208712152... \sim 1/5$

$\rho_{k,e} = -(3/2)\cdot(1-(7/3)^{1/2}) \sim 0.791287848... \sim 4/5$

The above numbers are the densities in modeling the $3\cdot n + 1$ with a Fibonacci sequence. From a numerical analysis for k = 32, we have [2]:

$\rho_{k,o} = 1174/5618 = 0.2089711641$

$\rho_{k,e} = 4444/5618 = 0.7910288359$

Thus:

$(\rho_{k,o}/\rho_{k,e})(k = 32) = 0.2089711641/0.7910288359 = 0.264176\ldots$

From the model:

$$\frac{\rho_{k,o}}{\rho_{k,e}} = \frac{\frac{(c-1)}{c}}{\frac{1}{c}} = c - 1 = \frac{\sqrt{\frac{7}{3}}-1}{2} = 0.2637626160\ldots$$

Remark:

For the $(3\cdot n + 1)/2$ where c is rational, c is given by c = 4/3 (Lagarias, 1985; Lagarias & Weiss, 1992; Lagarias, 2003; Lagarias, 2006), then $\rho_{k,o} = (c-1)/c = (1/3)/(4/3) = 1/4$ and $\rho_{k,e} = 1/c = 3/4$ so that here $\rho_{k,o}/\rho_{k,e} = 1/3 = 0.333..$ (Notice that, in this case, the Equation for c is given by $c^2 - c - (2/3)^2 = 0$ with a solution given by c = 4/3).

For the inverse orbits at depth k = 11 we have $N_{11} = 56$, $n_{k,o} = 15$, $n_{ke} = 41$; thus $\rho_{k,o}$ (k = 11) = 15/56 = = 0.267~1/4; $\rho_{k,e}$ (k = 11) = 41 /56 = 0.732 ~ 3/4 and from this calculation we find $(\rho_{k,o}/\rho_{k,e})$ (k = 11) = 15/41 = 0.365~1/3.

4.2.2. Density of the Odd and the Even Numbers on the Top of the Tree (Chalice) (Linear Density at a Fixed Level k of a Tree of Depth k)

The cardinality of the numbers at level k of the chalice of depth k (from above) is equal to the number of odd numbers in the full chalice and splits as $n_k = Ac^k = n_{k,o}' + n_{k,e}'$ where the symbol: ' refers to level k.

Now since $n_{k,e}' = n_{k-1,o} = A.c^{k-1}$ (for the chalice of depth k = 15, $n_{15} = 24$, $n_{k,e}' = 18$, $n_{k,o}' = 6$) we have that the linear density has the same value as above, i.e., from $n_{k,e}' = A \cdot c^{k-1}$, $n_{k,o}' = A \cdot c^k - A \cdot c^{k-1} = A \cdot c^{k-1} \cdot (c-1)$.

Thus, $\rho_{k,o}' = (c-1)/c$ and $\rho_{k,e}' = 1/c$, i.e., $\rho_{k,o}' / \rho_{k,e}' = c-1 = 0.26$.

For the $(3 \cdot n + 1)/2$, $\rho_{k,o}' / \rho_{k,e}' = c-1 = 1/3$.

A numerical experiment concerning the $3 \cdot n + 1$ formulation for values from n_{23} to n_{32} was presented by Merlini and Sala (1999, Table 2). The values of c provided by the stochastic model, using the results from n_{23} to n_{32} are satisfactory. For example:

$(n_{30} - n_{29})/n_{29} = 154/578 = 0.2664$, $(n_{31} - n_{30})/n_{30} = 192/732 = 0.2622... \sim (c-1) = 0.26376..$ for the $3 \cdot n + 1$ formulation.

4.2.3. Density (Linear) of the Odd and the Even Numbers (Ratio) in Some Special Inverse Orbit on the Tree

4.2.3.1. Choice of a Special Inverse Orbit Up to the Level k = 43 and More in the 3·n + 1 Formulation

It is interesting to construct a special inverse orbit on the chalice of depth k (for certain values of k) due to the probabilistic arguments present in the literature, concerning a geometrical mean that explains the number 3/4. This number is related to the relationship between the odd and even numbers appearing in a generic trajectory of length l(k) on the tree of depth k in the $(3 \cdot n + 1)/2$ (depicting an "exploding" factor of 3/2 for every odd in the $(3 \cdot n + 1)/2$ formulation and a contracting factor 1/2 for an even (n/2), essentially giving rise to the factor 3/4). This suggests that the Collatz conjecture will eventually be true (Lagarias, 1985). Here, our numerical "strategy" is to follow one (or a few) specific inverse orbits, that we have chosen as follows.

Strategy:

On the tree of the inverse orbits of various depths k, we look at an orbit having, locally, possibly a or the "high" number of odd. For example, on our tree of depth 15, for the 24 "local" orbits of length l = 15 starting at 1, the two inverse orbits (1, 2, 4, 8, 16, 5, 10, 20, 40, 13, 26, 52,17, 34,11, 22) and (1, 2, 4, 8, 16, 5, 10, 20, 40, 80, 160, 53, 106, 35, 70, 23) have five odd and eleven even numbers; in the $(3 \cdot n + 1)/2$ for each odd there is an even less (except for 1, See Eq.(8)) and the two orbits are given by: (1, 2, 4, 8, 5, 10, 20, 13, 26, 17,

11, 22) and (1, 2, 4, 8, 5, 10, 20, 40, 80, 53, 35, 23), of length l = 15-4 = 11 with five odd and seven even (l(22) = l(23) = 11 = $n_o + n_e - 1$ = 5 + 7-1).

See the tree of depth 15 in Figure 13 in the $3 \cdot n + 1$ formulation. Then, we continue and consider these two orbits (richer in odd) and their outcoming new trees, one starting at n = 22 and the other one at n = 23, and so on (local means some few steps in the inverse orbit: above 10 steps from n = 1, first up to k = 43 and then up to k = 118).

This construction was considered step by step up to the number 196609.

Step 1 of length 14: from 1 to 22-23.

Step 2 of length 11: from 22-23 to 492-493 (492 is a multiple of 3), l(493) = 48.

Step 3 of length 14: from 493 to 6231-6230 (6231 is a multiple of 3), l(6230) = 62.

Step 4 of length 36: from 6230 to 196608-196609 (196608 is a multiple of 3), l(196609) = 98.

This construction was considered first, step by step, up to the number n = 4153. If we obtain a number, multiple of 3, in a step, we do not follow their inverse orbit since one would obtain only even numbers, and thus the density (n_o/n_e) will, for a long orbit, $k \to \infty$, converge to zero.

Figure 14 shows the $n_o(k)$ and $n_e(k)$ plot in this special orbit for the 3n + 1. As can be seen, the quantity of even numbers is never less than that of the odd numbers, and the ratio approaches (for this special orbit) a possible maximum value of the "vertical" density given by:

$\rho_o = n_o/l = 1/3$ and $\rho_e = n_e/l = 2/3$ for the $3 \cdot n + 1$

and

$\rho_o = n_o/l = 1/2$ and $\rho_e = n_e/l = 1/2$ for the $(3 \cdot n + 1)/2$.

Acquired here by selecting a unique path among other options. (Note that for $3 \cdot n + 1$, 20/59 = 0.338~1/3, 39/59 = 0.661~2/3 while for the $(3 \cdot n + 1)/2$ 20/24 = 0.454 ~<1/2 and 24/44 = 0.545 > ~1/2). Below are the plots of l(k)/2, n_o, n_e in the $(3 \cdot n + 1)/2$ formulation.

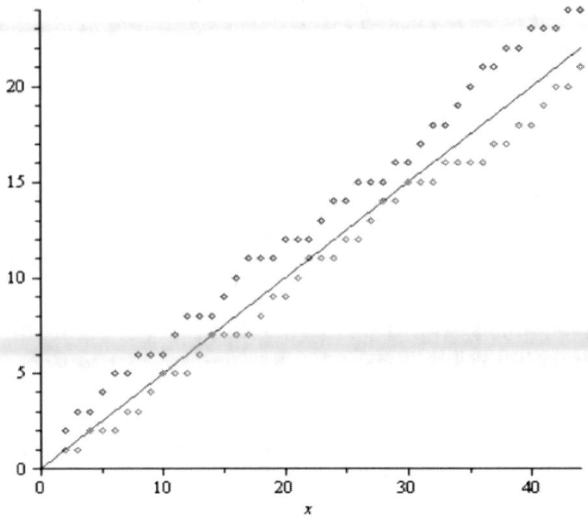

Figure 14. The straight line y = l(k)/2 = k/2 as a function of the depth k (l(k) = length of the inverse orbit) from 1 to 4153, $n_e(k)$ (points above the line) and $n_o(k)$ (below the line). The two sets are completely disjointed.

We may also follow the growth of the odd and even numbers with a plot of $(n_o/n_e)(k)$ in our orbit containing "many" odd (Figure 15).

We have (up to k = 43) for this orbit that $n_o \sim n_e$, but the experiment is too short and should be continued. Step by step, we have then reached the number n = 4418690. n = 4418690, l(n) = 118, l'(n) = 81, and the common number of odd is given by (118-81 + 1) = 38 = n_o, the number of even is n_e = 119-38 = 81 resp. n_e = 82-38 = 44.

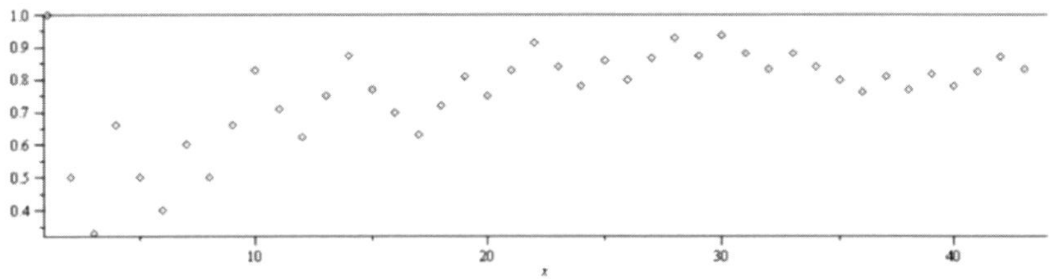

Figure 15. Plot of the constant 1 (the straight line) and $(n_o/n_e)(k)$ up to k = 43. In the $(3\cdot n + 1)/2$ formulation for the special orbit we have constructed. For this orbit, we have (up to k = 43) that $n_o \sim n_e$, but the experiment is too short and should be continued to approach the asymptotic value equal to 1 (see the conclusions).

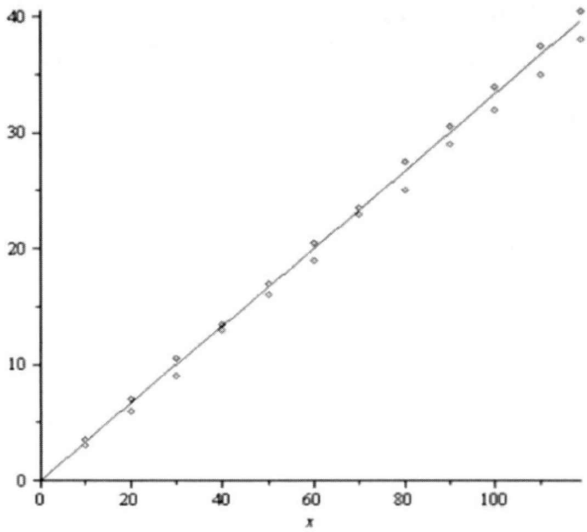

Figure 16. The straight line y = l(k)/3 = (k + 1)/3 as a function of the depth k (l(k) = length of the inverse orbit) n_e (k + 1)/2 (above the straight line) and n_o(k + 1), below the line, for the 3n + 1. The two sets of points are disjointed.

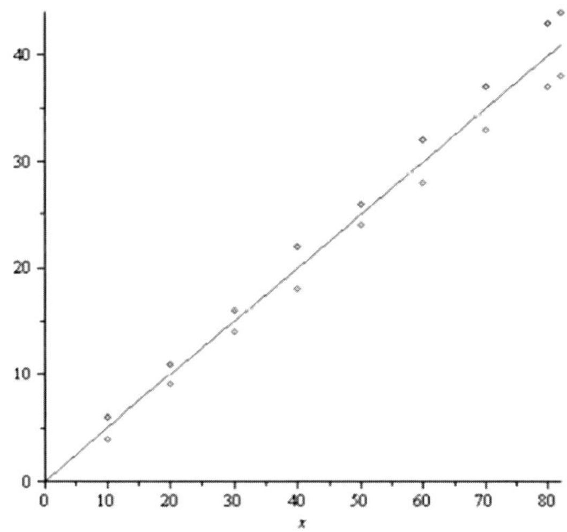

Figure 17. The straight line y = l(k)/2 = (k + 1)/2 as a function of the depth k (l(k) = length of the inverse orbit) n_e (k + 1)(above the line) and n_o(k + 1) below the line for the (3·n + 1)/2.

In Figures 16 and 17, we present the plots of n_e and n_o in the two formulations in steps of $\Delta k = 10$ as well as the ratios (n_o/n_e) $(k + 1)$, where k is the depth of the chalice of the inverse orbits from $n = 1$.

As it may be seen, the number of the even is here never or more minor than that of the odd, and the ratio approaches (for this unique orbit) a possible maximum value of the vertical density given by $\rho_o = n_o/l = 1/3$ and $\rho_e = n_e/l = 2/3$ for the $3 \cdot n + 1$ and $\rho_o = n_o/l = 1/2$ and $\rho_e = n_e/l = 1/2$ for the $(3 \cdot n + 1)/2$.

Remark:

We now present an argument supporting the potential validity of the Collatz conjecture alongside the established standard one.

Argument:

In an inverse orbit containing n_e even numbers and $n_o = \alpha \cdot n_e$ odd numbers, we take a factor $(3/x)$ for an odd (x unknown, finally set equal to $2-\varepsilon$, $\varepsilon>0$) and a factor $(1/x)$ for each even number (in the $(3 \cdot n + 1)/2$ formulation).

We then ask whether the upper bound of this process is given by a number less than one and then solve the following system of inequalities (let $x < 2$): $n \cdot (3/x)^{(\alpha . ne)} \cdot (1/x)^{(ne)} < 1$

$$(3 \cdot n + 1)/4 < n \cdot 3/x^2 = 3 \cdot n \cdot (1/(2-\varepsilon)^2) \sim 3 \cdot n/4 \cdot (1 + \varepsilon)$$

with $\varepsilon > 1/(3 \cdot n)$.

Then $\alpha < [\ (\log(2)-A)/\log(3/2)\ ] \cdot 1/(1 + B)$ where $A = (1/6n + \log(n)/n_e)$ and $B = 1/[6 \cdot n \cdot \log(3/2)]$.

For large n and long orbits, we have:

$$\alpha < [\ (\log(2)-\log(n)/n_e)/\log(3/2)\ ] \sim \log(2)/\log(3/2) - \delta = 1.709.. - \delta,$$

with $\delta > 0$ small.

For the $(3 \cdot n + 1)$ formulation we obtain

$$\alpha' < (\log(2)/\log(3)\ ) \sim 0.63093... -\delta.$$

That is, in all the orbits of the $(3 \cdot n + 1)/2$, the number of odd should be smaller than 1.709 times that of the number of the even ($n_{o/}/n_e < 1.709..$).

For certain small numbers with long orbits, α is around 4/3, smaller than 1.709.

Our orbit above up to k = 43 has $\alpha \sim 1$. As a further example, if we consider our saturation point n = l(n) = 73 (in the $(3\cdot n + 1)/2$ formulation), we have $n_o = 43$ and $n_e = 31$, and $\alpha = 43/3 = 1.387...<1.709..$

Now, for the $3\cdot n + 1$ formulation, $n_e' = n_e + \alpha\cdot n_e$ and $n_o' = n_o$ then, instead of α we have $2\cdot\alpha'$ and $\alpha' = \log(2)/\log(3) = \alpha/(1 + \alpha) = 0.63093..$

For n = 73, l(73) = 115, $n_o = 43$, $n_e = 116\text{-}43 = 73$ and $43/73 = 0.59 < 0.63093...$

In all the orbits of the $3\cdot n + 1$, the number of odd numbers should be smaller than 0.631. time that of the even numbers $n_o/n_e < 0.631$.

4.2.3.2. Study of Another Chosen Random Orbit

We now consider another case taken at random, i.e., the orbit of n = 16569146, which is derived from n = 1 in 10 steps.

Notice that l(16569146) = 200 and l ' = 132, i.e., the number of the odd is $n_o = 200\text{-}132 + 1 = 69$; $n_e = 200\text{-}69 + 1 = 132$ resp. $n_e = 132\text{-}69 + 1 = 64$ in the $(3\cdot n + 1)/2$ (Collatz Conjecture, 2024; Good Calculator. Collatz Conjecture $(3\cdot n + 1)$ Calculator, 2024).

We limit here to remark that this orbit is such that r = no/ne increases up to 1.3..., and at l(n) = 200, the value decreases to 1.078 in the $(3\cdot n + 1)/2$ case, not far from the unity as the special orbit of the first example c_1 (here r near 1, but from above). The plot of $r = n_o/n_e$ is given in Figure 18. Increasing by steps of 10, we reach n = 78550021 with r(240) = 72/68 = 1.058 < 1.078.

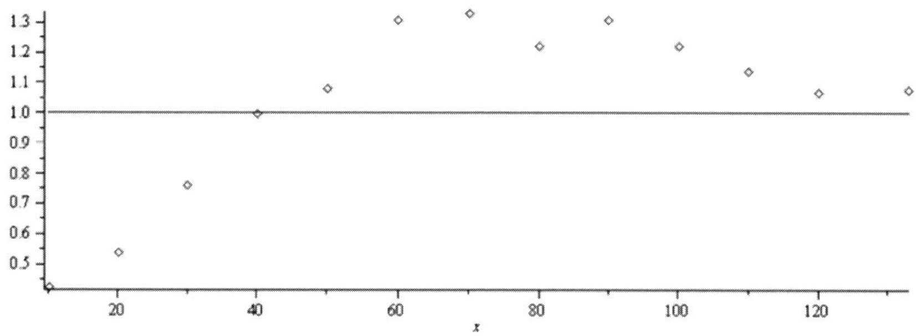

Figure 18. The plot of $r = (n_o/n_e)(k)$ in the $(3\cdot n + 1)/2$ formulation "approaching" r = 1 from above and the constant 1(the straight line).

4.2.3.3. Other Special Long Orbits (Enlarging the Tree of n = 703)

Now, for a typical number n of the table of Lagarias-Weiss, concerning a set of 10 numbers with a large total stopping time given in the Collatz Conjecture, six of them are multiples of 3. Thus, by increasing the inverse orbit by a finite amount of steps, one obtains only even numbers, α or α' are decreasing, as the length of the orbit increases, and for a very long orbit, the density of the odd to the even falls in these cases, to zero.

From the same table (Lagarias & Weiss, 1992, Table 2), the four numbers not multiple of 3 are: 703, 837799, 8400511, and 13371194527.

For 703, we have: l(703) = 108 = 62 + 46, for the (3·n + 1)/2 case, and no/ne = (63)/46 = 1.37...< 1.709 while for the 3·n + 1 case, where n_o' = 63, n_e' = 108 we have 63/108 = 0.583< 0.631.

We then limit the study of n = 703 to confirm α' < 0.631 and α <1.709.. for the most extended orbit starting from it, i.e., in steps reaching first the two numbers n = 1329072 and n = 1329080 where l(n) = 230, l'(n) = 149, no = 230- 149 + 1 = 81 n_e = 231-81 = 150 and n_e' = 150-81 = 69, the first number is a multiple of 3; ten steps more for the second number and we reach n = 6300823 (n = 240).

For this case (n = 6300823), we have checked that both α and α' are decreasing, possibly later approaching the value r ~1 (in the (3·n + 1)/2) and approximately ~ 0.5 (in the 3·n + 1) as the depth of the tree emerging from the second Lagarias-Weiss number, i.e., n = 703, increases progressively.

The above values satisfy, in any case, those emerging from the argument given above (1.7... and 0.630..). Table 1 (step Δk = 10 up to k + 1 = N = 150 and Δk = 6, 151-157).

Figure 19 illustrates the plot of r = (no/ne)(k) in the (3·n + 1)/2 formulation.

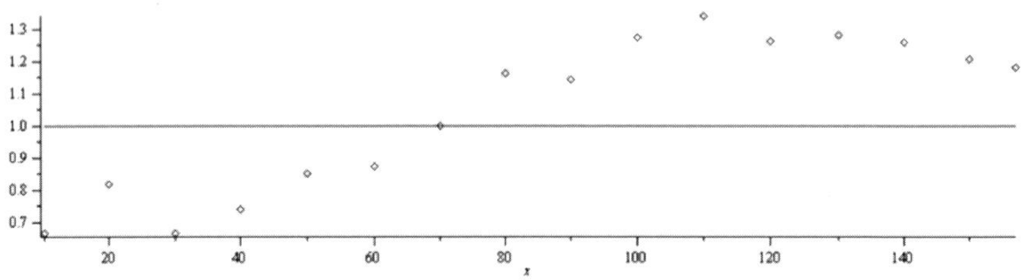

Figure 19. Plot of r = $(n_o/n_e)(k)$ in the (3·n + 1)/2 formulation "approaching" r = 1 from above and the constant 1 (the straight line), up to n = 6300822; at N = 156 + 1 = 157, r = 85/72 = 1.18.. at k = 156.

As noticed above, for n = 703, we have:

l(703) = 170, l'(703) = 108 number of odd: n_o = 170 -108 + 1 = 63; n_e = 109-63 = 46 in the (3·n + 1)/2.

Then, we have enlarged the tree of 703 up to the number n = 671329552 in step by 10, and the ratios are given for clarity in Table 1 (only the last step has Δk = 9 instead of Δk = 10).

Table 1. Total odd numbers up to n: 93 and total even numbers up to n:84 in the (3n + 1)/2 formulation

63/46 = 1.3695
66/53 = 1.2452
73/56 = 1.3035
77/62 = 1.2419
82/67 = 1.2238
86/73 = 1.1780
92/77 = 1.1948
94/84 = 1.1190

This computation "confirms" the decreasing of the ratio r = $(n_o/n_e)(k)$ at high values of k from above "toward" the value r = 1 (See conclusions).

4.2.3.4. An Orbit With r = 1 (a Multiple of 3) and Lim(r(k))→0 as k→∞

It concerns the number n = 1636057152, a multiple of 3; r = (no/ne)(k) is presented below in steps of 15. Notice l(n) = 217, l'(n) = 145, n_o = 217-145 + 1 = 73; n_e = 146-73 = 73 in the (3n + 1)/2 formulation; such n which (still falls in the orbit of n = 703) has exactly r = 1 with a maximum of r = 1.3.. and as other orbits, r fall first to 1 and then (since it is a multiple of 3) to r(k)→0 as k→∞. Figure 20 illustrates the plot of r(k + 1) for k + 1 = 15, 30, 45,..., 146.

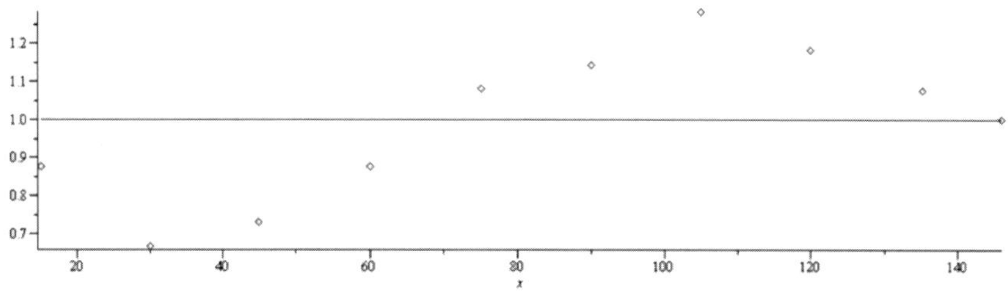

Figure 20. Plot of r (k + 1), k + 1 = 15, 30, 45,..., 146 for n given above (multiple of 3)·r(146) = 1.

Conclusion

Different approaches were used to analyze the 3·n + 1 problem (Lagarias, 1985; Lagarias & Weiss, 1992; Merlini, Sala & Sala, 1996; Merlini & Sala, 1999; Lagarias, 2003; Lagarias, 2006; Barina, 2021; Bishoff, 2021; Machado, Galhano & Cao Labora, 2021; Mohan Gurbaxani, 2021; Wegner, 2021; Merlini & Sala, 2022; Polli et al., 2024).

Recently, a connection between the Collatz conjecture and dynamical systems has been made.

In particular, the Collatz map is cited as a paradigmatic example of an arithmetic dynamical system. It can also be regarded as exhibiting key characteristics of an arithmetic statistical physics system, which accounts for its dynamic richness (Polli et al., 2024).

From our analytical and numerical study, we have checked that our weaker linear "bound" ($l(n) < n$, $n > n_o$) does not contradict the logarithmic one, i.e., $l(n) < c \cdot \log(n)$ ($n > n_s$) (which appears naturally in various analysis as well as in some stochastic models), and gives rise to the indication of the emergence of a saturation point in models of type 3·n + a, where a is odd.

For the 3·n + 1 model, we think that a closed cycle containing n odd does not exist except $C_{1,1}$, i.e., that the only solution of the set of Equations for a cycle containing n odd is given by:

$$\frac{(3 \cdot d_k + 1)}{2^{a_k}} = d_{k+1}, k = 1,..n-1$$
$$d_n = d_1$$

is $a_k = 2$ for each k and $d_k = 1$ for each k.

That is the "multicycle" 1→4→2→1→4→2→1→4→2→ ..., i.e., the cycle 1→4→2→1 repeated for n times as given by the above formula (see Figure 21). As for the open (vertical) orbits, where the "parameter":

$$r = \frac{n_o}{n_e}(k) \rightarrow \frac{1}{2}$$

As k→∞, even for the cycle 1→2→ 1.

We expect r→1 for the (3·n + 1)/2, as k→∞.

It would be fascinating to prove or disprove that as k →∞, r(k) = 1 in the (3·n + 1)/2.

For the number a not equal to 1, we know that other "small" cycles exist in addition to the cycle a→ 4a→ 2a→ a, i.e., the trees whose reunion contains all integers, not multiples of a.

An example with only an extra cycle is the case 3·n + 7 with the occurrence of $C_{2,7}$ (in addition to all the multiples of a = 7 if the Collatz conjecture for the 3·n + 1 is true).

For a = -1, i.e., 3·n-1, n positive, there are three cycles where we expect any integer to fall in one of the three cycles $C_{1,-1}$, $C_{2,-1}$, $C_{7,-1}$, i.e., containing one, two, and seven odd numbers.

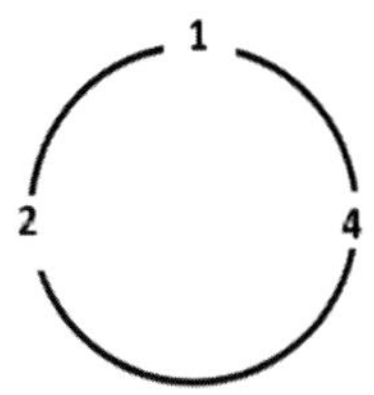

Figure 21. The "multicycle" 1-4-2.

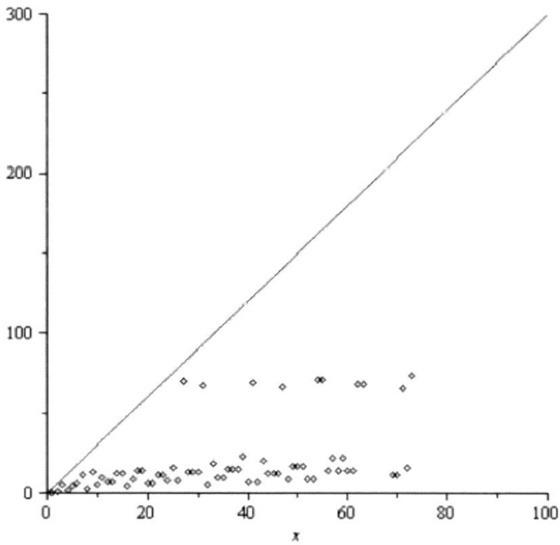

Figure 22. The plot of l(n) as a function of n in the (3·n + 1)/2 formulation and the function y = 3·n.

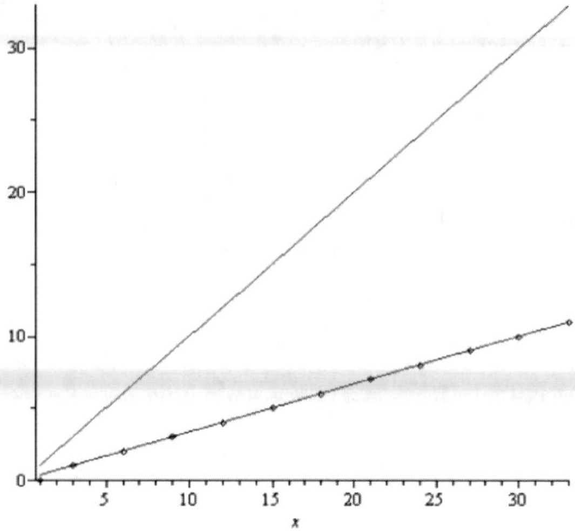

Figure 23. The plot of N(k) as a function of k (the depth of the tree), the functions y = k and y = k /3 in the (3·n + 1)/2 formulation.

Finally, it should be noted that with a weaker condition on the length of the orbits, the original saturation point at n = 118 for the 3·n + 1 and at n = 73 for the (3·n + 1)/2 disappears. This may be translated into the following new conjecture.

Conjecture:

For the (3·n + 1) formulation on a tree of depth k, there are, for each k, at least the first k/5 integers, and in the (3·n + 1)/2 formulation, at least the first k/3 integers. (In the first case, for each n, l(n) <5·n, and for the second case, l(n) < 3·n). See Figures 22 and 23.

References

Barina, D. (2021). Convergence verification of the Collatz problem. *The Journal of Supercomputing*, 77(3), 2681–2688.

Bishoff, M. (2024). The Simplest Math Problem Could Be Unsolvable. *Scientific American*, retrieved on June 2024 https://www.scientificamerican.com/article/the-simplest-math-problem-could-be-unsolvable/.

Collatz Conjecture on dCode.fr [online website], retrieved on April 2024, https://www.dcode.fr/collatz-conjecture.

Good Calculator. Collatz Conjecture (3n + 1) Calculator [online website], retrieved on January 2024 https://goodcalculators.com/collatz-conjecture-calculator/.

Lagarias, J. C. (1985). The 3x + 1 Problem and Its Generalizations. *The American mathematical monthly: the official journal of the Mathematical Association of America*, Vol. 92, Issue 1, 3 – 23.

Lagarias, J. C. (2003). The 3x + 1 problem: An annotated bibliography (1963–1999). *The ultimate challenge: the 3x*, 1, 267-341.

Lagarias, J. C. (2006). The 3x + 1 problem: An annotated bibliography, II (2000-2009). arXiv preprint math/0608208.

Lagarias, J. C., & Weiss, A. (1992). The 3x + 1 problem: two stochastic models. *The Annals of Applied Probability*, 229-261.

Machado, J. A. T., Galhano, A., & Cao Labora, D. (2021). A clustering perspective of the Collatz conjecture. *Mathematics*, 9(4), 314.

Merlini, D., & Sala, N. (2022). Saturation in the 3·n + 1 problem and a conjecture, *IOSR Journal of Mathematics* (IOSR-JM) e-ISSN: 2278-5728, p-ISSN: 2319-765X. Volumc 18, Issue 1, 01-10.

Merlini, D., & Sala, N. (1999). On the Fibonacci's Attractor and the Long Orbits in the 3n + 1 Problem. *International Journal of Chaos Theory and Applications,* Special Issue Volume 4, Number 2-3, 75 – 85.

Merlini, D., & Sala, N. (1999). Mappe discrete e numeri naturali: alcune considerazioni sul problema del 3n + 1. *Bollettino della Società Ticinese delle Scienze Naturali*, Vol. 87, n. 1-2, 33 – 37.

Merlini, D., Sala, M., & Sala, N. (1996). Il problema del 3n + 1 e la costante di Collatz, *Atti del convegno nazionale Mathesis:Cento anni di matematica, Fratelli Palombi Editore*, Rome, 331 – 336.

Mohan Gurbaxani, B. (2021). An Engineering and Statistical Look at the Collatz (3n + 1) Conjecture. arXiv e-prints, *arXiv-2103.*

Polli, J. G., Raposo, E. P., Viswanathan, G. M., & da Luz, M. G. (2024). Stochastic-like characteristics of arithmetic dynamical systems: the Collatz hailstone sequences. *Journal of Physics: Complexity*, 5, 1–22.

Wegner, F. (2021). The Collatz Problem generalized to 3x + k. arXiv preprint *arXiv:2101.08060.*

Chapter 5

The Butterfly Effect in the Chaotic Nonlinear Damped and Forced Oscillator

Javier Montenegro Joo*

Universidad Nacional Mayor de San Marcos, UNMSM, Lima, Perú
VirtualDynamics: Science and Engineering Virtual Labs, Lima, Perú

Abstract

This work reports the Butterfly Effect in the chaotic nonlinear damped and forced oscillator, which transitions to chaos employing cascades of period bifurcations. Altered chaotic events resulting from small changes in initial data taken as reference were obtained, and when compared to the original events, it was verified that the studied system experiences a Butterfly Effect. However, this does not manifest only as a final result; rather, the small changes in the initial data influence the evolution of the system, leading it down entirely different routes. It is observed that the cascades of period bifurcations in the altered events develop a notably different structure to that of the reference prototype events. It has also been observed that the evolution of the Lyapunov exponents during the flow of the altered systems follows remarkably different dynamics than that of the reference event. The more outstanding and enjoyable outcome of this investigation is that the number of periods the system oscillates with is raised when the initial input data is slightly increased. Additionally, some numerical data were collected during the computer simulation of the studied chaotic events, and these reflect significant differences when compared with the data from the prototypes.

---

* Corresponding Author's Email: jmontenegroj@unmsm.edu.pe

In: Complexity and Fractality in Human Change and Beyond
Editor: Nicoletta Sala
ISBN: 979-8-89530-575-1

Keywords: butterfly effect, chaos, nonlinear dynamics, nonlinear damped and forced oscillator, differential equation, computer simulation, sensitivity, initial conditions

1. Introduction

The Butterfly Effect establishes that the simple folding of a butterfly's wings in some place, which is interpreted as a small event, may trigger a huge happening like a tornado in another faraway place (Lorentz, 1993). Speaking of huge events, notice that this may not necessarily be a tornado or a tragedy; the Butterfly effect may also imply that the original minor alteration of initial conditions; might avoid the huge effect. This consequence is known as sensitivity to initial conditions, and, from the mathematical perspective, it means that if those conditions are infinitesimally changed in a system known to be chaotic, then the system will evolve towards a significantly different state.

In other words, the Butterfly Effect may be regarded as a huge and disproportionate reaction to a relatively small stimulus. In this way, the Butterfly Effect refers to the enormous -not necessarily negative- consequences produced by hardly noticeable changes in the initial conditions. At this point, "small" and "enormous" are relative terms because what is termed "hardly noticeable" when compared with what is deemed "enormous" may be gigantic in some other circumstances. Thus, a new question arises at the time of computationally assessing the Butterfly effect: what to consider a rather small change in the initial conditions and what to consider an enormous consequence.

However, there still is another point to observe here and although it is very small or hardly noticeable, it does not necessarily mean insignificant, for when it happens that the folding of the butterfly's wings is capable of either generating or avoiding a tornado, that hardly noticeable event may be instead considered a highly significant event, the moral of this is that small does not necessarily imply insignificant.

Since not all chaotic systems behave continuously chaotic, in specific experiments based on computer simulation, it has been observed that an infinitesimal change in the initial conditions might take the system from regular, i.e., smooth non-chaotic dynamics, to chaotic dynamics and vice versa.

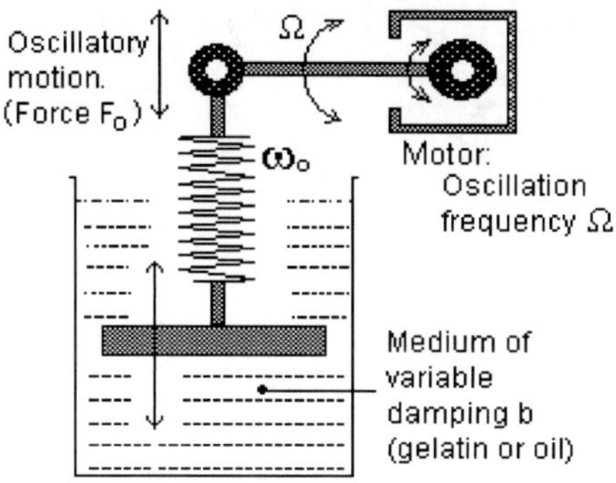

Figure 1. The nonlinear damped and forced oscillator, NLDFO.

In situations like this, an infinitesimal alteration of the initial conditions of a chaotic system may cause it to behave non-chaotically.

In Figure 1, the nonlinear damped and forced oscillator (NLDFO) is sketched; a motor forces a piece attached to a spring to oscillate while it remains immersed in gelatin or oil. As the system oscillates, the temperature of the liquid changes, and the medium varies its density. Eventually, there is a conflict between the natural frequency of the spring and that of the applied force; the system starts to oscillate weirdly and adopts an unpredictable, chaotic behavior. The dynamics of this system are represented mathematically by the following differential equation:

$$\frac{d^2x}{dt^2} + 2b\frac{dx}{dt} + \omega_o^2 \, Sin\, x = F_o \, Sin\, \Omega\, t \qquad (1)$$

In this equation, b is the damping offered by the liquid the oscillator is immersed in, ω_o is the natural frequency of the oscillator, F_o is the applied external oscillating force whose frequency is omega (Ω).

In this research, instead of lowering the density (viscosity, damping) of the medium, it is kept constant, but the applied force F_o is gradually but slowly increased. Equation (1) is an inhomogeneous ordinary differential equation of second order (Steiner, 2003; Kreyszig, 1964), and it is one of those equations where obtaining an analytical solution, this is, an explicit mathematical expression as a solution, is rather complex, if not impossible, it is due to this reason that it is solved in a computer by the so-called numerical methods and, the most common approach is the Runge-Kutta algorithm (Weber & Arfken, 2003; Riley, Hobson & Bence, 2006). In this investigation, the system modeled by equation (1) is simulated and assessed along six-million-time steps (time units).

2. The Experiment

The chaotic event identified by this researcher as #1 in the nonlinear damped and forced oscillator (NLDFO) is obtained by simulation of equation (1) with the following parameters:

$$b = 1.0, \omega_o = 2.0450, \Omega = 1.47020, Initial\ F_o = 8.850,$$

$$\Delta F = 0.00000010, x_o = 0, v_o = 0, \Delta t = 0.0250$$

In this research, this is called the reference chaotic event, whose period bifurcation cascade is shown in Figure 2 and whose initial conditions will be slightly altered subsequent to see their effect on the system's dynamics. The top of Figure 2 exhibits the peaks of the displacement x(t) in the oscillations of the NLDFO, and its bottom displays the corresponding valleys. Since the period of an oscillation may be assessed from the horizontal distance of points with the same phase in the plotting of x(t), Figure 2 represents the oscillation periods as time passes. In this case, it is evident that there are several period bifurcations and that regions of strict chaotic behavior (dotted zones) are sporadically interrupted by zones of relative tranquility where the system oscillates alternating among a few well-defined periods. Notice in Figure 2 that the chaotic event starts and finishes with the system vibrating within a single period. Be aware that the maximum number of periods the system oscillates within this experiment is five.

Figure 3 exhibits the first altered chaotic event; this period bifurcation cascade results after introducing minor modifications to the initial frequencies in the reference event shown in Figure 2; in this case, the modified parameters are $\omega_o = 2.050$ and $\Omega = 1.47050$.

Comparing the bifurcation cascades in Figures 2 and 3, careful comparison reveals that there are some notable differences between the two events, although the difference between these cascades is not so apparent at first glance, which could be because chaotic events in the NLDFO begin with the generation of periodic bifurcation cascades and end with collapsing cascades, although not necessarily the same (Montenegro, 2018). A more evident modification due to a slight alteration of original data is shown in Figure 4, which displays the second altered chaotic event; in this case, the modified parameters are $\omega_o = 2.0655$ and $\Omega = 1.48700$, which represents an alteration of a little more than 1% in the original frequencies (Figure 2). It is evident in Figure 4 that, this time, the evolution of the chaotic event is significantly different from those appearing in Figures 2 and 3, which display the reference and first altered event, respectively.

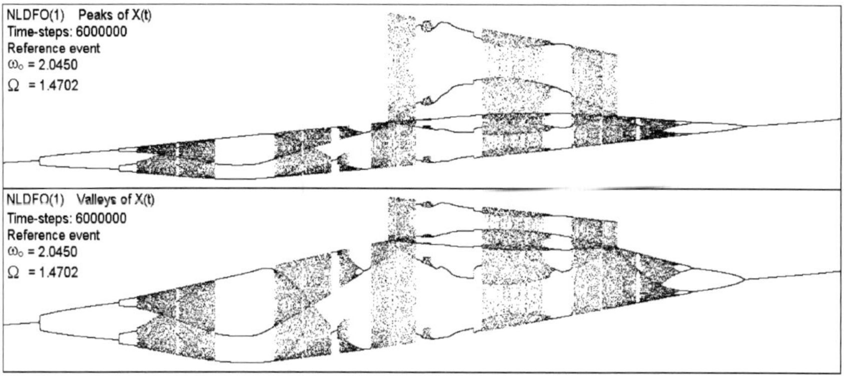

Figure 2. Period bifurcation cascade of chaotic event 1 in the NLDFO. This is considered the reference event in this research.

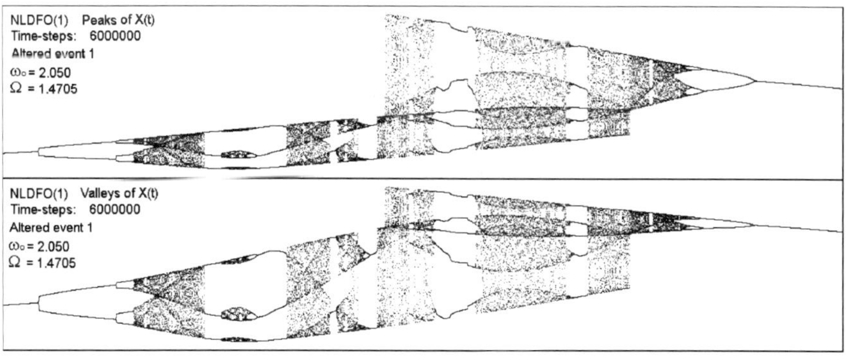

Figure 3. Period bifurcation cascade of the first alteration of the chaotic NLDFO(1). In this case, $\omega_o = 2.0500$ and $\Omega = 1.4705$. Notice that some new bifurcations have shown up in the first period-3 region.

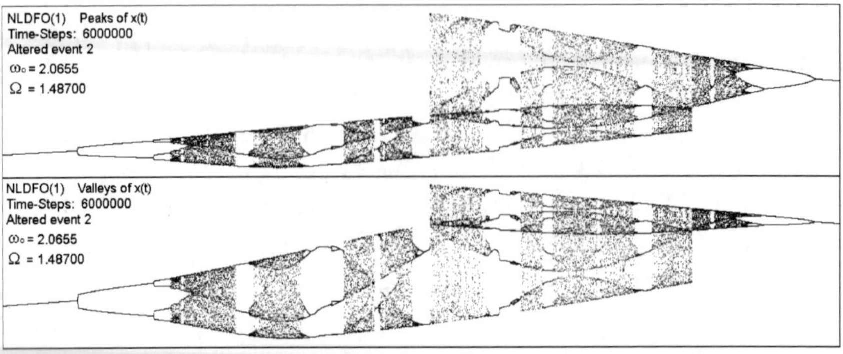

Figure 4. Period bifurcation cascade of the second alteration of chaotic event NLDFO(1). In this case $\omega_o = 2.0655$ and $\Omega = 1.48700$.

In the Butterfly effect, it is expected that a slight change in the initial data will be expected to cause the system to have vastly different results. These results are not necessarily greater nor smaller, just exceedingly different, however, if changes are gradual, it is common sense to expect a certain tendency in the behavior of the system, and this is precisely what is made in this research; this is, to introduce small but gradual changes in the input parameters, so that it is expected that the output is quite different but gradual.

3. Comparing the Three Events Studied in This Research

In the virtual lab developed to accomplish the experiments reported here, it is possible to detect the coordinates of the points where the bifurcations appearing in Figures 2, 3, and 4 start. To achieve this, the cursor of the mouse is positioned at the chosen point in the cascade, and it is found out that the first bifurcations in these three cascades start at 258317 time-steps in the reference event, at 253620 time-steps in the first altered event and at 526027 time-steps in the second altered event.

These different positions of the start of the first bifurcations are not evident at simple sight. By comparing the reference event in Figure 2 with the first altered event in Figure 3, it can be seen that some new bifurcations have arisen at the first period-3 region at the left side of the reference event. In Figure 4, displaying the second altered event, these new bifurcations have practically turned a region of relative tranquility where the oscillator was alternating among three periods into a fully chaotic stage.

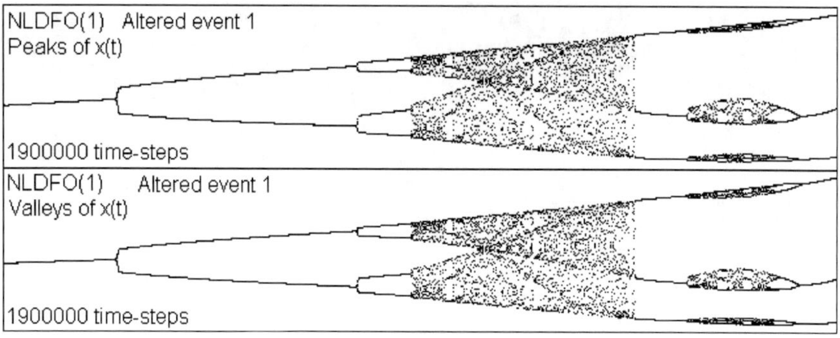

Figure 5. Amplification of the region in the first altered event where new bifurcations (appearing here at the far right) have appeared as a consequence of the initial data alteration.

Figure 5 displays an enlargement of the zone in the reference chaotic event (Figure 2) that was oscillating while alternating among three periods, then -as it is evident in this figure, in the first altered event, some new bifurcations appeared in this zone, and in the second altered event (Figure 4) this zone appears chaotic. When Figure 5 is watched on a computer scrollable screen, it is evident that a 9-period stage appears among the new small bifurcations.

This means that one of the more significant changes produced by the initial data alteration in the reference event is the increment in the number of periods the system alternates among while oscillating.

4. Maximum Number of Periods in the Cascades

Observe that the 9 periods the system alternates among in Figure 5 are not detected at a glance in Figure 3; however, in the period bifurcation cascade of the second altered event shown in Figure 4, there is a region where the vibrator -at simple sight- oscillates alternating among 9 periods, in this occasion neither an especial visual effort nor an amplification is necessary.

In Figure 6, after oscillating alternating among 5 periods in 'a,' the vibrator enters during a short time in strict chaos in 'b,' and this stage is interrupted by an oscillation among 9 periods in 'c,' sometime later, the system returns to its chaotic vibration in 'd,' which is again interrupted by a short time oscillation among 5 periods in 'e.'

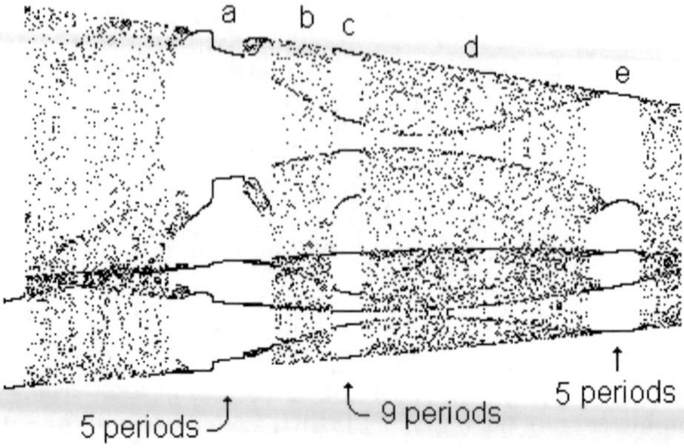

Figure 6. Nine periods were detected at the simple sight in the second altered event.

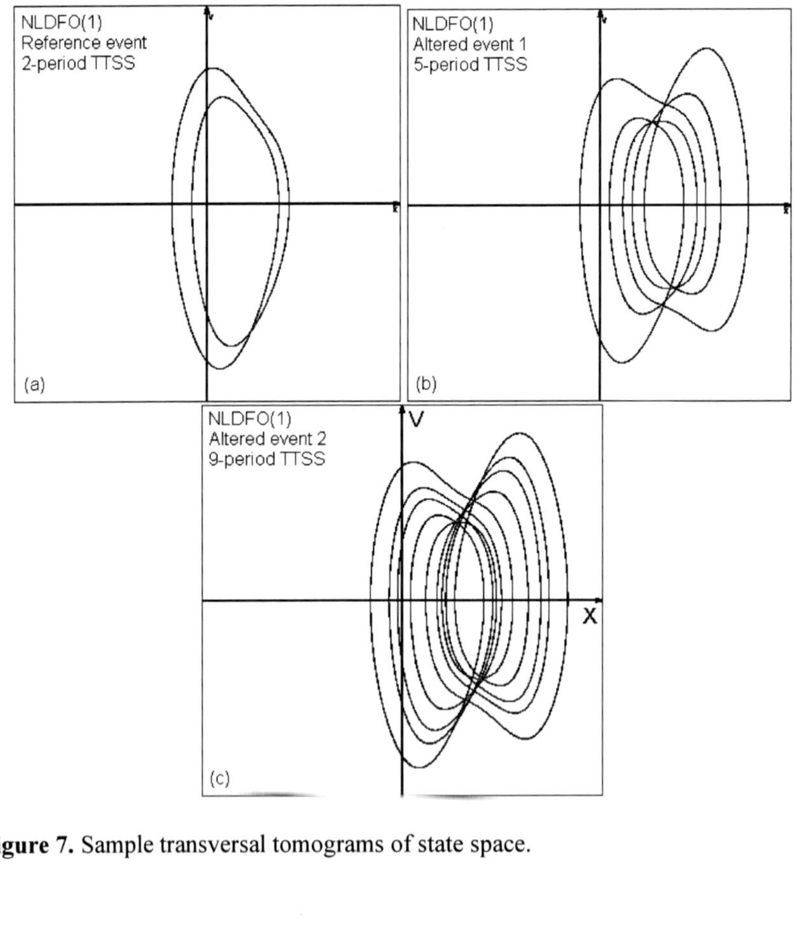

Figure 7. Sample transversal tomograms of state space.

Figure 7 presents Poincaré transversal tomograms of state space from the three period-bifurcation cascades studied in this investigation. A two-period zone of state space is exposed in (a), whereas image (b) displays a five-period zone; periods like these appear at a glance in the three events analyzed in this work.

A nine-period zone appears in (c), and this is evidently present in the second altered event, though with some visual effort, it is also detectable in the first altered event (Figure 3). Notice that when the vibrator is in a stage like the one depicted in (a), it is alternating between two frequencies while it oscillates, when the vibrator is in (b), it alternates among five frequencies and, when the vibrator is in a stage like (c), it vibrates alternating among 9 frequencies.

5. Other Details Observed in the Resulting Period Bifurcation Cascades

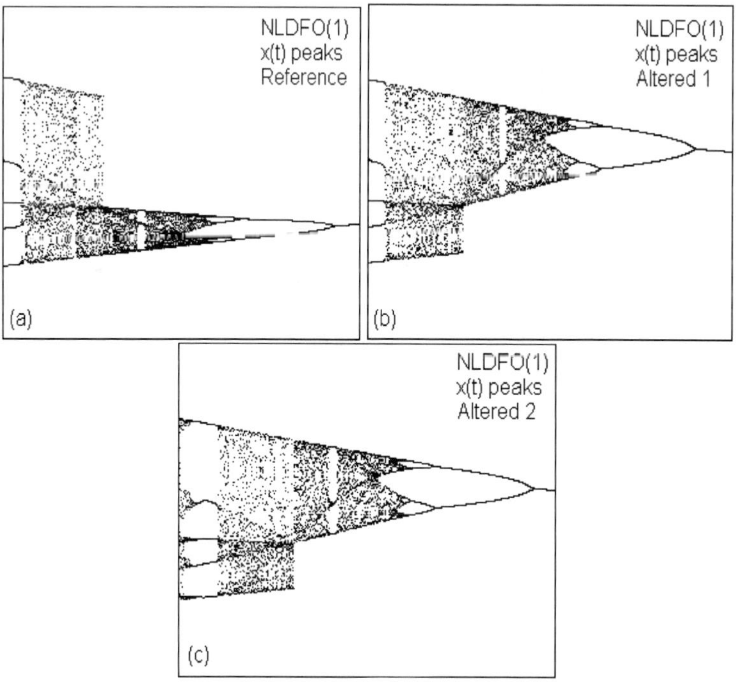

Figure 8. Comparison of the final stage of the period-bifurcation cascades (peaks) in the three events considered in this research.

As can be seen in Figure 8, the final stages of the period bifurcation cascades in the reference, and both altered events are different; in the reference cascade, the final stage -shown in (a)- appears at the lower wing, whereas in both altered events -displayed in (b) and (c)- the final stages of the bifurcations appear at the upper wings of the cascades. The three sections exhibited in this figure start after a 5-period zone, and an additional detail is that these sections span the same time length, but these start and finish at different instants.

Concerning the start of the period cascades appearing in Figures 2, 3, and 4, the first bifurcation starts almost at the same time in both; the reference and the first altered event; however, this first bifurcation is retarded in the second altered event. In the three generated chaotic events, the complexity of the vibrations increases from the reference event to the first and then to the second altered event, which means that the increase of initial data in the studied events gets reflected in the increasing complexity of the resulting events.

It is worthwhile to mention that when watching the period-bifurcation cascades on a computer scrollable screen, many more details arise than when watching the images in a printed report.

6. Numerical Results of the Simulations

Concerning the numerical data obtained in the simulations carried out in this research, Table 1 exhibits some results from the simulations of chaotic events in the NLDFO(1), data from the reference event and the two altered events are shown. In this research, the input values of both the natural frequency ω_o of the oscillator, and the frequency Ω of the applied oscillating force are in two opportunities, slightly changed to higher values, which must indicate a certain tendency in the output parameters; this effect is evident when going over data in the table.

6.1. Analysis of Data in Table 1

In this research, both frequencies, natural and applied, were gradually and little by little risen in each simulation with equation (1), and their values appear in the first two rows of Table (1); the other rows are the output of the simulation and, the right-most column in Table 1 indicates -using arrows- the tendency observed in the data in the same row. By reviewing data in Table 1, it is evident that the highest peaks and deepest valleys in the curve of displacement versus time x(t) result in higher and deeper, respectively. This is confirmed by the progressively increasing distances from the highest peaks to the deepest valleys; likewise, the number of peaks and valleys in the curve of x(t) is steadily increasing.

The number of expansions in the state space is associated with the number of positive Lyapunov exponents, whereas the quantity of contractions is related to the number of negative exponents. Results in the table suggest that more expansions are occurring and the number of contractions seems to decrease, since the average Lyapunov exponent indicates the overall behavior of the system, then in the studied cases, it can be inferred that in general, as the values of both, ω_o *and* Ω are increased, the state space widens. It can also be seen in the table that the Kolmogorov entropy of the studied cases tends to increase, which is associated with the general widening of state space.

The value of the maximum Lyapunov exponent does not respect any gradual increment in this experiment; however, the value of the minimum Lyapunov exponent does follow a tendency, and the same happens with the average Lyapunov exponents.

Table 1. Data detected during the evolution of the studied chaotic events

Chaotic event	Reference	Altered 1	Altered 2	
ω_o	2.0450	2.050	2.06600	↑
Ω	1.47020	1.47050	1.48700	↑
Max x(t) = highest peak	8.26392004	8.31632473	8.38885658	↑
Min x(t) = deepest valley	-2.53335841	-2.58220377	-2.63589939	↓
Distance Max x(t) – Min x(t)	10.797278451	10.898528499	11.024755966	↑
N x(t) peaks = N x(t) valleys	35098	35105	35499	↑
Max Lyapunov Exp: λ_{max}	3.4083 s^{-1}	3.4124 s^{-1}	3.4078 s^{-1}	↕
Min Lyapunov Exp: λ_{min}	-2.3417 s^{-1}	-2.3515 s^{-1}	-2.3815 s^{-1}	↓
Ave Lyapunov Exp: $< \lambda >$	1.2113 s^{-1}	1.2125 s^{-1}	1.2232 s^{-1}	↑
Number of $\lambda > 0$	4881004	4882050	4892718	↑
Number of $\lambda < 0$	1118996	1117950	1107282	↓
Min Pos Lyapunov Exp	2,969E-008	4,731E-007	2,621E-006	↑
Orbit Divergence = Max_d - do	8,894E-009	8,905E-009	8,893E-009	↕
Kolmogorov Entropy K	8425903.0683	8435373.5460	8511176.6166	↑
Ln(K)	15.9468	15.9479	15.9568	↑

The number of positive Lyapunov exponents gradually increases in the three studied events, which implies an increased number of expansions of the orbit in state space. The number of negative Lyapunov exponents shows a gradual decrease, indicating that the number of state space orbit contractions also decreases.

If the objective of this research would be the study of the evolution of the parameters in Table 1 as the initial data in the oscillator are gradually changed, the number of experiments should be much larger, which means that the number of columns in the table should be higher; however, the data currently appearing in the table suggests us roughly what to expect from such research.

7. The Lyapunov Exponent Evolution in the NLDFO(1)

The Lyapunov exponent (Wolf et al., 1985; Farmer et al., 1983), which is represented by λ is a measure of orbit deformation in state space, and it is assessed by means of

$$\lambda = \frac{1}{t} ln\left[\frac{d_t}{d_o}\right] \tag{2}$$

In this equation, d_o is a very small change introduced in x(t) before computing each time-step, d_t is the outcome of the computation, and t is the elapsed time for that computation. If there is no change after the computation, then $d_t = d_o$ *and* $\lambda = 0$. Should it result that $d_t > d_o$ and consequently $\lambda > 0$, then the orbit has expanded, and, when $d_t < d_o$ and therefore $\lambda < 0$, then the orbit has experienced a contraction (Sprott, 2003; Rosenstein et al., 1993). Concerning the number of $\lambda > 0$, this indicates the number of orbit expansions in state space and, similarly, the number of $\lambda < 0$ indicate the number of orbit contractions (De Souza-Machado et al., 1990).

When there is no change in the orbit after introducing d_o, the Lyapunov exponent results $\lambda = 0$, but the number of $\lambda = 0$ is impossible to count because λ is a real number, and for it to be precisely zero in the computer, it is necessary to have all its digits equal to zero, which is almost impossible as a result of a complicated computation; in this case, an approximation must be made (Montenegro, 2020).

Since the simulations in this research run along 6 million time steps, the graphs rendering the evolution of the Lyapunov exponents must be distributed to fit the limited space and be understood; otherwise, the graphs would display a dark spot impossible to appreciate It is due to this condition that the three curves shown next exhibit the evolution -every 50000-time steps- of the Lyapunov exponent for the reference chaotic event (Figure 9), for the first altered-data chaotic event (Figure 10) and, for the second altered-data chaotic event (Figure 11). Figure 9 shows the evolution of the Lyapunov exponent in the chaotic event in the NLDFO taken as a reference; in this case, the values for the frequencies were $\omega_o = 2.0450$ and $\Omega = 1.4702$.

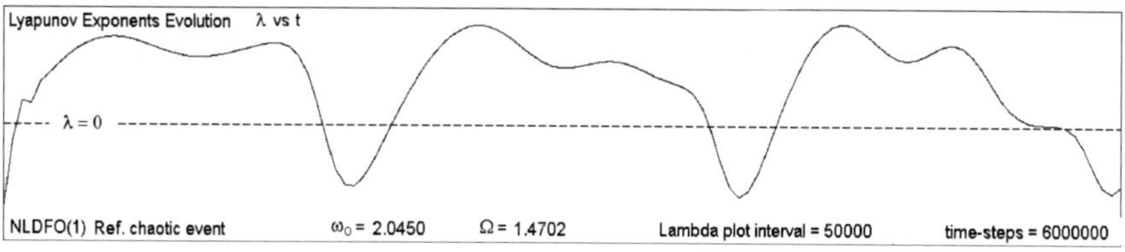

Figure 9. Lyapunov exponent evolution in the chaotic NLDFO(1), the reference event.

Then, as a first data alteration, small changes were introduced for both frequencies, the parameters being in this case $\omega_o = 2.050$ and $\Omega = 1.4705$. The results of this change may be seen in Figure 10.

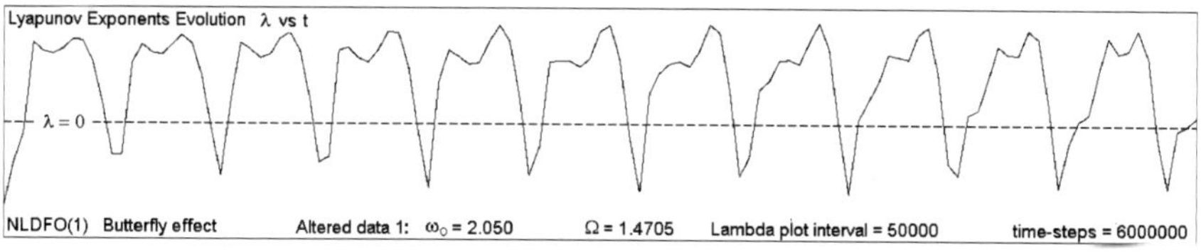

Figure 10. Lyapunov exponent evolution of the NLDFO(1) with the first altered data.

Finally, a second small data alteration was introduced, and this time, the parameters were $\omega_o = 2.0655$ and $\Omega = 1.4870$, and the results of this change are shown in Figure 11.

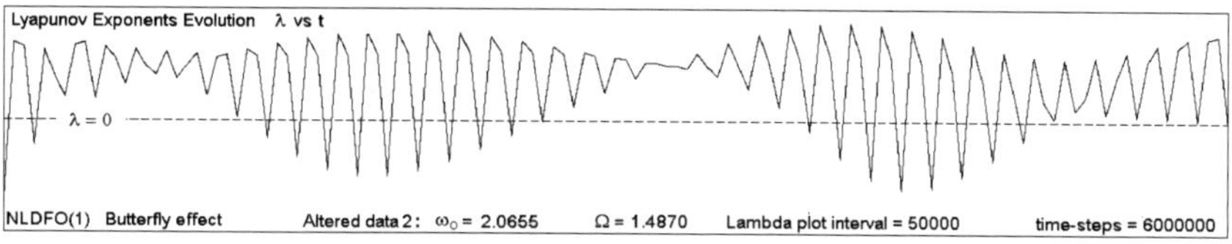

Figure 11. Lyapunov exponent evolution of the NLDFO(1) with the second altered data.

By comparing Figures 9, 10, and 11, it is clearly seen that the small data alterations to the reference event, whose Lyapunov exponent evolution is shown in Figure 9, guide the system through quite different dynamics. These results confirm that chaotic systems are strongly sensitive to initial conditions so that small alterations of primary sceneries steer the evolution of the system through completely different behavior; this is known as the Butterfly Effect.

8. The Total State Space of the Simulations

Figure 12 displays the total state space, from beginning to end, for the reference chaotic event in the NLDFO. In the three studied cases - reference, altered 1, and altered 2- the bi-dimensional version of state space is the same, which may be justified by the fact that in this work, the complete chaotic events, from the beginning through the end, are studied and plotted. If there is any difference in the dynamics of the system -let´s say, in the middle of the event- this gets overshadowed in the graph of the state space by the subsequent plotting, which is almost equal in all cases; in order to verify this, give a glance to the bifurcation cascades displayed in Figures 2, 3 and 4, in these three cases the chaotic events start by generating bifurcations and finish by collapsing bifurcations, though not necessarily the same ones.

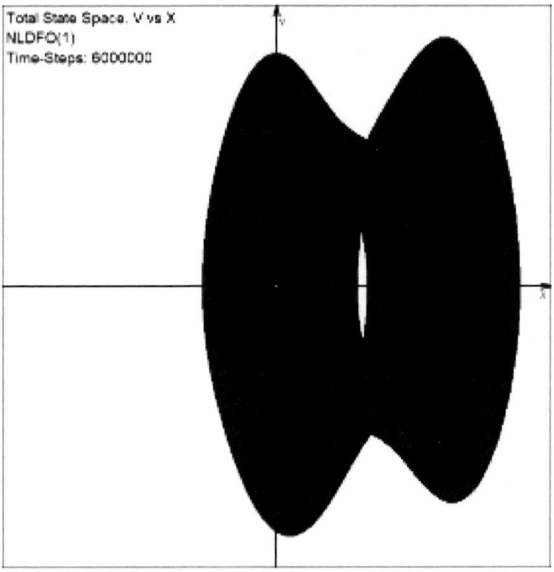

Figure 12. In the three studied cases, there is no difference in the graphs of the total state space, it is due to this reason that only the reference state space is shown.

9. Maps of Return

In the early studies of Chaos, the primary reference was the Logistic Equation, which is a mathematical model of population evolution and whose Map of Return is a parabola-shaped distribution of dots that is obtained by the plotting of an oscillation amplitude versus its preceding amplitude (Gonzales-Miranda, 2004), this is, the graph of X(n + 1) versus X(n). For a constant oscillation where the amplitudes are uniform, the map of return is a single dot and, in the case of a damped oscillator where amplitudes gradually decrease, the map is a set of dots aligned towards the origin of coordinates in the space of X(n + 1) versus X(n), complicated dynamics like that of nonlinear and chaotic vibrators generate more complicated maps of return (Montenegro, 2018) (Montenegro, 2019).

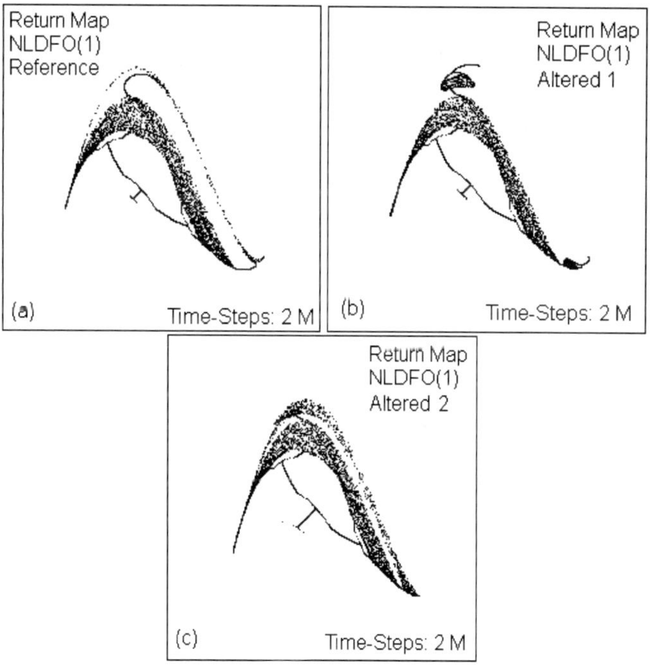

Figure 13. Maps of Return obtained from the first 2 million peaks of the amplitudes. The maps for the reference event and the two altered events are shown.

Figure 13 shows the maps of return for the reference event and the first and second altered events in this research; these maps were obtained for the first 2-million-time steps. As it can be seen, the dynamics of the studied events are rather complicated and different in the three events.

In general, the maps of return of non-regular dynamics generate difficult to interpret graphics, making it complicated to get useful information from them; maybe for this reason, the use of these maps is not so common.

10. Dual Maps of Return

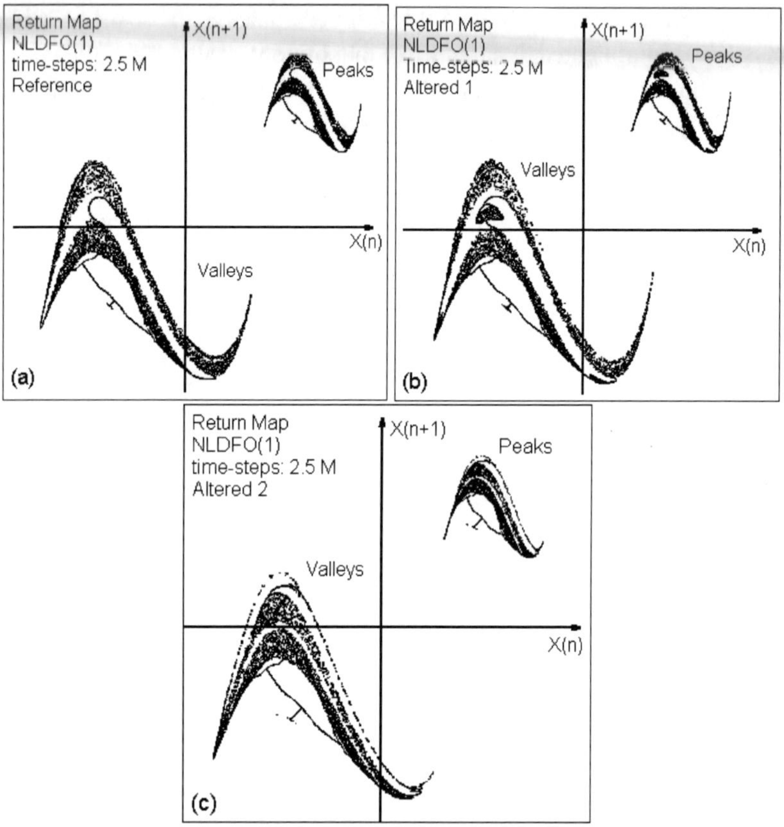

Figure 14. Maps of Return of the first 2.5 million time-steps. Some differences appear in the three studied events. (a) the reference event, (b) the first altered event, and (c) the second altered event.

The period bifurcation cascades shown in Figures 2, 3, and 4 make evident the fact that the extremes of the amplitudes are not symmetric, this is, the distribution of peaks and valleys of displacement in chaotic oscillators is rather far from being uniform, for this reason, when dealing with this type of vibrations it is better to generate dual maps of return namely, those including simultaneously maps of peaks and maps of valleys and, this is what is exhibited in Figure 14 for the first 2.5 million time-steps, of the simulations.

It can be seen in the dual maps of return that the peaks of the oscillations studied are always on the positive side of the oscillation, whereas the valleys of the oscillation are on the negative side of the equilibrium position, but occasionally, they invade the positive side. The here exhibited maps of return have some differences which reflect the different evolution of the studied events. Notice in these images that the maps of return of the studied events have all self-similarity and fractal structure.

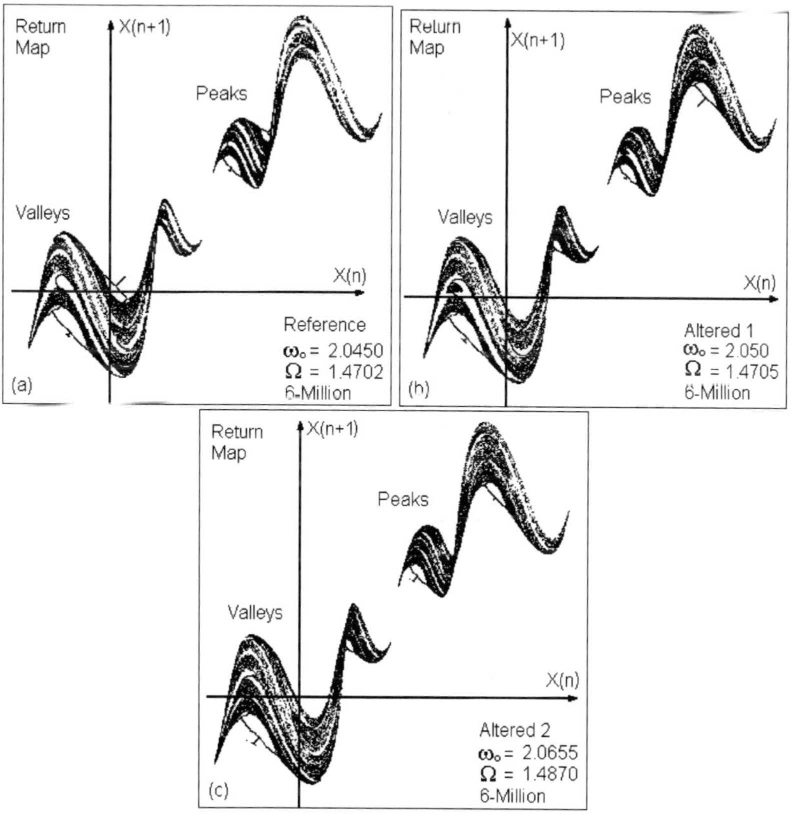

Figure 15. Dual maps of return of the three chaotic events of the NLDFO(1) along six million time-steps: (a) The reference event (b) the altered event 1, and (c) the altered event 2.

Figure 14 displays the dual maps of return of the first 2.5 million time-steps in the studied events, whereas Figure 15 shows the dual maps of return of the complete events, this is, 6 million time-steps and, when these two graphs are contrasted it is evident that some details occurring during the evolution of the events have been covered in the maps of return of the complete events, this justifies the analysis of partial maps of return.

11. The Butterfly Effect in Real Life

When all the above exposure about the Butterfly effect is generalized to real life, it makes us remember that some huge wars of the past and also of the present were supposed to be rather short or were started as military campaigns of relatively small consequences; however, they unchained all the demons of the hell. On a much minor scale, everyone has some examples of the butterfly effect in their lives; hence, in general, it seems to be that this effect is highly disseminated in everyday life; then a quick inference at this point is that either the Butterfly effect is not limited to chaotic systems or real life is a chaotic system. After a little thinking, it can be noticed that, in fact, both chaotic oscillators and real-life follow nonlinear dynamics and, hence, the butterfly effect must be common to both of them.

An important detail worth mentioning at this point is that when working with computational algorithms like Cellular Automata (Wolfram, 1994) and other procedures used in rather complex computer simulations, like stochastic aggregations (Vicsek, 1992), genetic algorithms (Michalewicz, 1992), Neural networks, etc., the sense that real life resembles a huge computer simulation arises as a natural consequence. Some scientists support the notion that real life is a gigantic and sophisticated computer simulation. The butterfly effect from chaos theory adds further support to this way of thinking.

Conclusion

The Butterfly effect in the nonlinear damped and forced oscillator (NLDFO) has been studied. It has been verified that small alterations in the initial conditions of a reference chaotic event guide the system through significant changes as exploration events occur, the most notable effect being an increase in the detectable number of periods the vibrator alternates while it oscillates.

To achieve this investigation, the oscillator's natural frequency and the applied force's frequency were altered. The research was carried out by implementing slight but gradual changes in the input parameters of a chaotic event taken as a reference. The results of these gradual changes are different but not always necessarily gradual. In this research, the notably different results due to infinitesimally changed input data are mainly appreciated in the evolution of the Lyapunov exponents of the studied cases.

The differences in the period bifurcation cascades of the reference and altered events are not always evident to the naked eye, and something similar happens with the maps of return. The fact that chaotic events in the NLDFO start by generating period bifurcation cascades and finish those events by collapsing bifurcation cascades tends to reduce some of the average output parameters, giving the impression that the changes in the outputs are not so significant as they would look like if the chaotic events were finished right after a change is observed in the altered events.

References

Farmer, J. D., Ott, E., & Yorke, J. A. (1983). Nonlinear phenomena, The dimension of chaotic attractors. *Physica D.*

Gonzalez-Miranda, J. M. (2004). *Synchronization and Control of Chaos*. London: Imperial College Press.

Kreyszig, E. (1964). *Advanced Engineering Mathematics*. New York: John Wiley & Sons.

Lorentz, E. (1993). *The Essence of Chaos*. Seattle: University of Washington Press.

Michalewicz, Z. (1992). *Genetic Algorithms + Data Structures = Evolution Programs*, New York: Springer-Verlag.

Montenegro, J. J. (2018). *Anatomía del Caos: Estudio del Caos en modelos matemáticos, basado en Física Computacional y Simulacional*, Editorial Académica Española.

Montenegro, J. J. (2018). Tomographies of the State-Space of the Chaotic NLDFO, *Chaos and Complexity Letters* 12(1), 71-82.

Montenegro, J. J. (2019). Analyzing chaotic dynamics by means of state space tomographies and dual maps of return, *Chaos and Complexity Letters* 13(1), 29-43.

Montenegro, J. J. (2020). Computational procedure to assess Lyapunov exponents as well as associated indicators in a chaotic event, *Chaos and Complexity Letters* 14(1), 33-44.

Riley, K. F., Hobson, M. P., & Bence, S. J. (2006). *Mathematical Methods for Physics and Engineering*, Cambridge University Press, New York, USA.

Rosenstein, M. T, Collins, J. J., & De Luca, C. J. (1993). A practical method for calculating the largest Lyapunov exponents from small data sets, *Physica D*, 65.

Sprott, J. C. (2003). *Chaos and Time-Series Analysis*. New York: Oxford University Press.

Steiner, E. (2003). *The Chemistry Maths Book*. New York: Oxford University Press.

Vicsek, T. (1992). *Fractal Growth Phenomena.* London: World Scientific Publishing.

Weber, H. J., & Arfken, G. B. (2003). *Essential Mathematical Methods for Physicists*, California: Academic Press.

Wolf, A., Swift, J. B., Swinney, H. L., & Vastano, J. A. (1985). Determining Lyapunov exponents from a time series. *Physica 16D.*

Wolfram, S. (1994). *Cellular Automata and Complexity.* Boulder, Colorado: Westview Press.

Chapter 6

Complexity and Fractality in Architecture: A Brief Analysis

Nicoletta Sala[1,*]
and Gabriele Cappellato[2]
[1]Institute for Scientific and Interdisciplinary Studies, ISSI, Locarno, Switzerland
Research Institute in Arithmetic Physics, IRFIA, Locarno, Switzerland
Institute for Complexity Studies, Rome, Italy
[2]Emeritus Professor Academy of Architecture, Mendrisio, Switzerland

Abstract

Complexity is a theory that has its roots in the late nineteenth-century works by the mathematical physicist Henri Poincaré (1854-1912) and in those of the first half of the 20th century by mathematicians and physicists such as Hadamard (1865-1963), Lyapunov (1857-1918), Schrödinger (1887-1961), Kolmogorov (1903-1887), Andronov (1901-1952).

Complexity is also a property of a real-world system that manifests in the inability of any one formalism to adequately capture all its properties.

Fractality is the quality of being fractal. The term derives from the word "fractal," a neologism coined by Benoit Mandelbrot (1924-2010) in 1970. It derives from the Latin word "fractus," meaning "broken, fractured." These adjectives describe the fractal objects (irregular and jagged).

Complexity and fractality have applications in many disciplines, including engineering, computer science, medicine, biology, and physics. However, can they also be applied in architecture?

* Corresponding Author's Email: nicolettasalavb@gmail.com

In: Complexity and Fractality in Human Change and Beyond
Editor: Nicoletta Sala
ISBN: 979-8-89530-575-1

Modern architects try to apply non-linear properties to create new buildings using modern materials or to understand the problems connected to network organizations and urban growth.
This chapter describes an analysis that studies complexity and fractality applied in architecture, presenting examples from different cultures across the centuries.

Keywords: box-counting dimension, complexity, fractality, fractal geometry, non-linear architecture, self-similarity

1. Introduction

The word "complexity" comes from the Latin verb "complector," which means to enclose, hold tightly, and metaphorically embrace and bring together under a single thought and a single denomination. Other meanings in the Latin classics are bond, connection, and concatenation.

In Encyclopedia Britannica, Casti (2023) defines complexity as: "Complexity, a scientific theory which asserts that some systems display behavioral phenomena that are completely inexplicable by any conventional analysis of the system's constituent parts."

Sporns (2007) defines complexity as "… the degree to which components engage in organized structured interactions."

In *Complexity and Adaptation* (1986), Huberman and Hogg propose that complexity is a mixture of order and disorder (Figure 1).

Complexity is connected to chaos, revealing a hidden fractal order underlying all seemingly chaotic events. It can occur in natural and man-made systems, meteorological systems, human beings, and social structures. Complex dynamical systems can be small or large; small and large components coexist cooperatively in some complex systems.

In *Aesthetic Measure* (1933), the American mathematician George Birkhoff (1884-1944) proposed a measure of beauty connected to complexity. He presented a mathematical theory of aesthetics based on a simple relationship:

$$M = \frac{O}{C} \qquad (1)$$

M stands for "aesthetic measure" (or beauty), O for order, and C for complexity. This measure suggests that beauty has something to do with order and complexity.

Birkhoff's theory attempts to apply complexity to aesthetic measures. In the past, different mathematical and geometrical elements were employed to seek an aesthetic sense and perfection in the arts and architecture, such as Euclidean geometry, symmetry, the golden ratio, and Fibonacci's sequence (Ghyka, 1977; Blackwell, 1984; Hargittai & Hargittai, 1996; Dunlap, 1998; Sala & Cappellato, 2003; Sala & Cappellato, 2019).

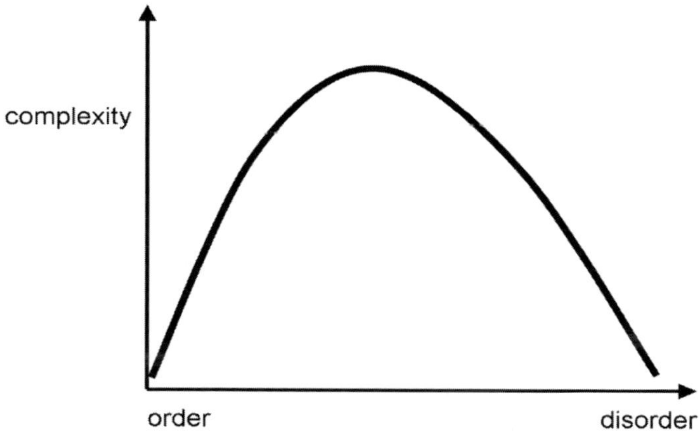

Figure 1. Complexity as a measure of order and disorder.

2. Complexity and Fractality in Architecture

We can also observe the architecture from a different point of view. For example, we can find the symmetry or the golden section, or fractal and complex components that are present in the buildings (Venturi, 1992; Eaton, 1998; Jencks, 1998; Sala, 2002; Sala & Cappellato, 2004; Sala, 2006; Sala, 2007; Sala & Cappellato, 2019). In his book entitled *Complexity and Contradiction in Architecture* (1992), Robert Venturi affirms:

> "The recognition of complexity in architecture does not negate what Louis Kahn has called 'the desire for simplicity.' But aesthetic simplicity which is a satisfaction to the mind derives, when valid and profound, from inner complexity" (Venturi, 1992, p. 17).

Complexity and fractality appear in architecture in different ways, for example, for reproducing natural patterns and natural shapes or for breaking symmetry (Jencks, 1998; Sala, 2002; Sala, 2003; Sala & Cappellato, 2004; Sala, 2006; Sala, 2007; Sala, 2012; Sala & Cappellato, 2018; Sala, 2018; Sala & Cappellato, 2019).

Our complex analysis of architecture has been divided into two parts:

- on a small-scale analysis (e.g., to determine the complex components present in the building shape);
- on a large-scale analysis (e.g., to study urban growth and urban development or to analyze landscape organization).

In this chapter, we will present a small-scale analysis, which we have organized into three parts:

- the complexity of the buildings' shape (e.g., the complex surfaces and the complex textures in the works of Franck O. Gehry);
- the fractality of the buildings' shape (across the different cultures);
- the box-counting dimension of a design is used to determine its degree of complexity (for example, when applied to Mesoamerican architecture and modern architecture).

3. The Complexity of the Buildings' Shape

In ancient oriental architecture, we can observe intricate shapes that remind us of the complex components. An example is Prambanan (or Candi Prambanan or Candi Rara Jonggrang). It is a 9th-century complex of Hindu temples. This archeological site is located approximately 20 kilometers northeast of Yogyakarta on the boundary between Central Java and Yogyakarta provinces (Indonesia). The complex of temples is dedicated to the Trimurti (the Trinity of Hinduism). The Trimurti is the supreme expression of divinity in Hinduism, and it comprises the Creator, Brahma, the Preserver, Vishnu, and the Destroyer, Shiva. Figure 2 shows a view of the three temples.

Complex decorations are present in the Muqarnas, known in Iberian architecture as Mocárabe. It is a three-dimensional architectural decorative element in Islamic architecture in which rows or tiers of niche-like elements are projected over others below. This decorative element flourished in its most complete form mainly during the Islamic period and is most pervasively used in semi-domes and domes. One of the key features of this fascinating element is its intricate geometry, which remains a mystery and is a subject that requires deeper study by artists and art historians. Figure 3a shows Decorative muqarnas vaulting in the entrance to the *Shah Mosque* (1611-1629) in Isfahan (Iran), which can remind the vault of a cave with stalactites (Figure 3b).

Figure 2. The *Temples of Prambanan* represent an example of Hindu temples, where the shape reminds the complexity.

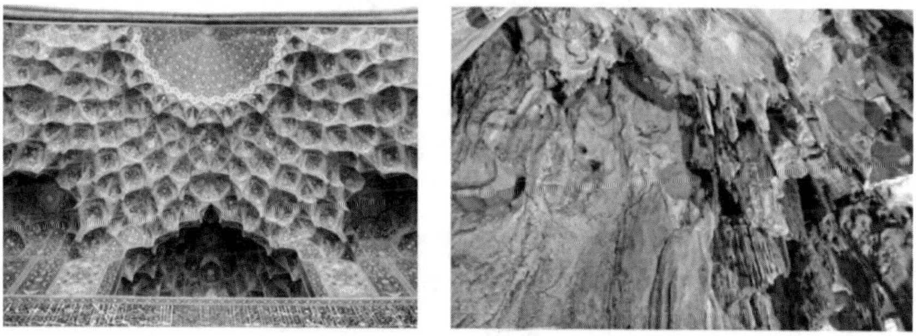

a) b)

Figure 3. Decorative muqarnas vaulting in the entrance to the *Shah Mosque*, Isfahan, Iran (a). Vault of a cave with stalactites (b).

Recently, a new type of architecture emerged from the complexity of non-linear design (Jencks, 1998). To research the complexity of the buildings, we can observe the complex textures (e.g., in Gehry's works) and the intricate surfaces (e.g., in Gehry's or Portoghesi's projects) derived from the observation of nature and natural phenomena.

The Canadian-born American architect Frank O. Gehry is one of today's most inventive and pioneering architects. He has used the complexity and the fractality in his recent works. Since his fish lamps in 1983, Gehry has applied the property of self-similarity to realize the complex textures of his buildings. This is evident when the metal shingle, repeated on different scales, covers the Vitra Headquarters (1989-1992) and the Guggenheim Museum, Bilbao (1992-1997). Figure 4 shows a snake's skin, where the complexity and the self-similarity are evident. Figure 5 illustrates a portion of the Guggenheim Museum's skin. The analogy with nature is amazing.

Figure 4. The skin of the snakes has self-similar textures.

Figure 5. Gehry used self-similarity to identify the skin of the *Guggenheim Museum*, Bilbao (1992-1997).

Figure 6. Gehry Buildings of Dusseldorf Harbor (Dusseldorf, Germany).

Figure 7. The wave motion is an example of a complex system.

The Gehry Buildings of Dusseldorf Harbor (Dusseldorf, Germany) mimic the chaotic movements connected to the gas or liquid motion (Figure 6). That could be compared with smoothed surfaces inspired by the wave motion as a metaphor for the sea (Figure 7).

Austrian architect Gunther Domenig realized some buildings that recall complex and fractal shapes; an example is Steinhaus (1998-2008) at Lake Ossiach (Carinthia, Austria). This house has an open outward planimetry in which distorted volumes of cement or steel are inserted, starting from a glazed cylinder that directly emerges from the subsoil.

In his project, Domenig captures a moment of initial chaos. Therefore, his work is not ended and, at first glance, appears "destroyed." The structure grew year by year, piece by piece, following an ever-evolving set of sketches and technical drawings.

Italian architect Paolo Portoghesi, inspired by the chaotic movements associated with gas or liquid motion, designed the Hotel Savoia (1992-1996) in Rimini, Italy, featuring smoothed surfaces that reflect wave motion as a metaphor for the sea. The hotel is located on the Adriatic. (Portoghesi, 1999). Portoghesi, Domenig, Gehry, and other contemporary architects, have consciously developed a non-linear architecture. Two examples of non-linear architecture are in Figures 8a and 8b.

a)

b)

Figure 8. Modern architecture finds inspiration from non-linear and complex shapes.

4. Fractality in Architecture

Fractal geometry describes fractal objects as mathematical and complex. These objects are created using simple iterative or recursive algorithms.

The fractal objects have a unique property; it is called "self-similarity." Self-similarity is the characteristic of certain natural and mathematically produced forms that exhibit features that appear identical or similar at different scales of magnification.

Mandelbrot, in his book *The Fractal Geometry of Nature* (1982), observed that this property is ubiquitous in the natural world and the human body. It is possible to observe the presence of fractal geometry using two different points of view: temporal fractals and spatial fractals.

Unconscious fractal components are present in different cultures and different styles. Recently, we discovered fractal components in architecture as a deliberate design choice. This paragraph presents examples of fractal architectures, organizing their presentations in continents (Africa, Asia, Europe, America, and Oceania).

a)

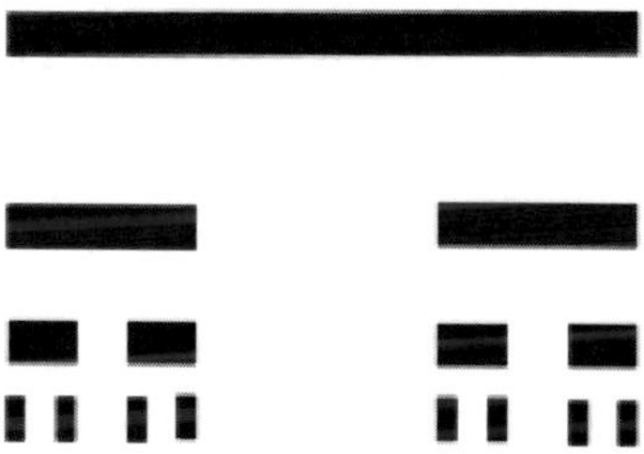

b)

Figure 9. The scheme of an Egyptian capital (a) shows an interesting similitude with a Cantor set (b).

In Africa, many fractal references exist. In ancient Egypt, the Pyramids are examples of using polyhedral forms and scale reduction. Another interesting example is the capital of an Egyptian temple column. A particular scheme of an Egyptian capital is shown in Figure 9a. The Ancient Egyptian cosmogony often used the white lotus flower to represent the universe's development. The lotus' corolla is organized in petals within petals within other petals; in this way, the lotus represents the cosmos on smaller and smaller scales. This is a clear example of self-similarity. A similar cosmos organization is present in Hindu philosophy. We can compare the stylized lotus petals and the similarity of a representation to the first three iterations of Cantor's set (Figures 9a and 9b). This capital is probably one of the oldest fractal artifacts.

In the book *African Fractals, Modern Computing and Indigenous Design* (1999), American scholar Ron Eglash describes some examples of fractals in art, architecture, and African design. Eglash emphasizes that African architecture mirrors a community's social and religious structure. The architectural examples presented by the author show that fractal features are a direct consequence of some structural or organizational features of a settlement. In some cases, they represent a religious or social hierarchy. Eglash claims, from a political perspective, that European urban planners consider African settlements as big villages and not real cities, as it does not use, as in Europe, Euclidean geometry for street tracing and urban organization, but rather complex shapes that resemble fractal geometry. He says that fractal architecture was used as colonial proof of primitivism, and the chaos present in African architecture was used as a symbol of European fears over social chaos (1999, p. 196). Eglash observed three interesting fractal organizations in the African settlements. They are circular fractals, rectangular fractals, and branching fractals. An example of circular fractals is demonstrated by Ba-Ila, situated in Southern Zambia (Africa). Every house features a circular shape, is enclosed by a cattle fence, and includes a primary entrance and an exit. Adjacent to the main hall, small structures are serving as food dispensers. Each house is encompassed by a circular fence, defining the settlement perimeter. Continuing around the circle, the buildings, maintaining their uniform shape, gradually increase in size, leading to the largest one that serves as the residence of the head of this community and is situated directly across from the main entrance. In front of this house, there is the chief's house. Ba-Ila represents an example of a self-similar settlement based on circular fractals. An example of rectangular fractals is in Logone-Birni (Cameroon).

Another example of African fractal architecture is the *Great Mosque of Djenné* (Djenné, Mali). It is an interesting example of the Sudano-Sahelian architectural style. The current structure of this mosque dates back to the first decade of the 20th century (1907), but the first mosque on the site was built around the 13th century. It is built using adobo (or adobe). Adobo is a mixture of clay, sand, and often organic material, such as sun-dried straw, used by many populations at all times to realize bricks. Figure 10 shows an old view of the Great Mosque from the northeast.

Figure 10. View of the Great Mosque from the northeast as it looked in 1910 (From Félix Dubois' *Notre beau Niger*).

a) b)

Figure 11. *Temple of Kukulcan* (c. 8th – 12th century), also known as *El Castillo* (Chichen Itzá, Mexico) (a) and *Temple 1* of Tikal (Mexico) have fractal structures (b).

In America, fractal components existed in ancient civilizations. In particular, Mesoamerican civilizations built numerous administrative and ceremonial centers and erected numerous monuments (such as pyramids and tombs) that reflected astronomical knowledge and expertise in numeration and calendars. In Mesoamerican architecture, we can find the fractality. The *Temple of Kukulcan* (c. 8th -12th century), also known as *El Castillo* (Spanish for *"the Castle"*), is located on the site of Chichen Itzá (Mexico). It is a

Mesoamerican step-pyramid that dominates the center of the Chichen Itzá archaeological site in the Mexican state of Yucatán. It shows fractal components (Figure 11a) in its plan organization. This characteristic is also present in the *Temple of Great Jaguar* (*Temple I*) and the Temples of Tikal (Mexico) Figure 11b.

More recently, the American architect Frank Lloyd Wright (1867-1959), a key figure in modern architecture, incorporated fractal geometry into some of his projects, employing a deliberate fractal design. For example, in the *Palmer House* (Ann Arbor, Michigan, 1950-1951) (Eaton, 1998). The fractality is also present in the *Marin County Civic Center* (San Rafael, 1957) and the *Annunciation Greek Orthodox Church* (Wauwatosa, Wisconsin, 1956-1961). In *Palmer House*, he used self-similar equilateral triangles to organize the project. At the same time, in *Marin County Civic Center* and the *Annunciation Greek Orthodox Church,* cycloids are present on four different scales of size (Sala & Cappellato, 2019). Frank Owen Gehry often uses fractal geometry to highlight the rounded shapes and create a plastic dynamism, as in his *Walt Disney Concert Hall* (1999-2003) in Los Angeles (Figure 12).

The Polish-American architect Daniel Libeskind was fascinated by fractal geometry and self-similarity. In the *Art Museum* (Denver) and *Victoria & Albert Museum* (London), he uses them both to create light effects and to highlight the symmetry breaking.

The Spanish architect and structural engineer Santiago Calatrava, at the *Westfield World Trade Center* located inside the *World Trade Center Transportation Hub* (New York), is influenced by natural fractal forms (Figure 13).

In Asia, fractal components are present in Hindu architecture. In the Hindu experiences, the idea of the sacred mountain is represented in the cosmic mountain named Meru (with five peaks in Hindu, Jain, and Buddhist cosmology). Mount Meru is the mythological center or navel of the universe. The fractal structure of the mountains is well known and discussed by scholars, and it is often observable using a "fractal point of view." Thus, we can observe that Indian and Southeast Asian temples and monuments have a fractal structure because they found inspiration from nature, mainly from Mount Meru, and its fractality. The temples' towers are encircled by smaller towers, continuing for seven levels or more. In these instances, the abundance of towers on various scales signifies a characteristic of the Hindu pantheons, a form of self-similarity in their architectural composition.

Figure 12. *Walt Disney Concert Hall* (1999-2003). This Gehry's building shows fractal components.

Figure 13. *Westfield World Trade Center* is located inside the World Trade Center Transportation Hub by Calatrava.

The American scholar William Jackson asserts that the Hindu temple usually encompasses a variety of ideas (Jackson, 2002, p. 3). Hindu traditional architecture has more symbolic meanings than the architecture of other cultures (for example, European or American cultures). Hindu architecture is articulated, and its ideal forms suggest the infinite rising levels of existence and consciousness. Figure 14 shows an example of a Hindu temple with a self-similar shape.

Figure 15 shows another example of Hindu fractal architecture: the *Mandara Giri Semeru Agung Temple* (*Pura Mandara Giri Semeru Agung,* Java, Indonesia). This recent Hindu temple was finished in 1992, and it is located at Senduro, Lumajang Regency, on the slope of Gunung Semeru, the highest peak in Java. Figure 16 shows that the *Mother Temple* of Besakih, or Pura Besakih, in the village of Besakih on the slopes of Mount Agung in eastern Bali, Indonesia, is the most important, largest, and holiest temple of the Hindu religion in Bali. The architecture of the Buddhist temples recalls fractal shapes. In particular, the property of self-similarity. Figure 17 shows a view of *Wat Luang Phor Toh,* a modern and popular Buddhist temple in Sikhio town (Korat, Thailand).

Figure 14. An example of Hindu self-similar architecture. The main shape is repeated in different scales of representation.

In the Asian continent, Islamic architecture also employs self-similarity, particularly in arches and domes that replicate their shape at different scales.

The *Mausoleum* of Sitt Zubaida (1179-1225) in Baghdad, probably made by the Caliph an-Nassir on his mother's tomb, has as its highlights a conical dome where overlapping niches are becoming smaller and smaller. This dome type was common in hot and dry regions such as Lower Iraq.

Mughal architecture is an architectural style developed by the Mughals in the 16th -17th centuries during the ever-changing extent of their empire in Medieval India. It is a fusion of Persian, Islamic, Turkic, and South Asian architecture. Mughal buildings have slender minarets at the corners, large vaulted gateways, delicate ornamentation, and large bulbous domes of different scales as a type of self-similarity. Examples of this architectural style exist in India, Afghanistan, Bangladesh, and Pakistan. The *Taj Mahal* (Agra, India) (1632 – 1648) is the most well-known example of Mughal architecture. Its name was derived from "Mumtaz Mahal," the favorite wife. Self-similarity is present in this funeral monument in the arches' shape and the cupolas' shape repeated in four scales (Figure 18).

Figure 19 shows another example of Mughal architecture: *Humayun's Tomb* (1557). It is the tomb of the Mughal emperor Mirza Nasir al-Din Muhammad, commonly known as Humayun, located in Delhi, India. This tomb shows self-similarity in the arches and the domes.

Figure 15. *Mandara Giri Semeru Agung Temple* (Senduro Village, East Java, Indonesia) has self-similar components.

Figure 16. *Mother Temple* of Besakih, or Pura Besakih, in the village of Besakih on the slopes of Mount Agung in Eastern Bali, Indonesia.

Figure 17. *Wat Luang Phor Toh* is a Buddhist temple in Sikhio town (Korat, Thailand).

In Chinese and Japanese architecture, self-similarity can be found in the roofs of wooden pagodas (two examples are shown in Figure 20). Their simple original structure consisted of a base on which a cusp was raised, which protruded upwards to support a series of overhead roofs. Thus, the roofs multiply, overlapping with a decreasing dimension, creating a prospective effect. The central plant of the pagodas usually has a polygonal shape.

Figure 18. The *Taj Mahal* (1632-1648) exemplifies Mughal architecture. Self-similarity can be seen in the arches and domes.

Figure 19. *Humayun's Tomb* (1557) shows self-similarity in the domes and the arches.

In modern Chinese architecture, an example of a fractal is the 244-meter-tall *Olympic Park Observation Tower* in Olympic Green, Beijing, which was designed to resemble a fractal tree. Another interesting Chinese building is the *Beijing National Stadium* (2003-2008), also known as the *Bird's Nest,* for its complex shape (Figure 21). The Swiss architects Jacques Herzog and Pierre de Meuron have designed this stadium in collaboration with other important architecture and engineering firms.

Figure 20. The simple structure of the pagodas shows the self-similarity in the roof organization.

Figure 21. Fractal Chinese architecture: *Beijing National Stadium* (2003-2008), also known as *Bird's Nest.*

Figure 22. Fractal organization in the buildings of Venice (Italy).

In Europe, Gothic and Baroque architecture had complex decorations. Frequently, their cathedrals had self-similarity expressed on several levels. Gothic is a style developed in northern France that spread throughout Europe between the 12th and 16th centuries. "Gothic" was a term first used during the later Renaissance by the Italian artist Giorgio Vasari (1511-1574) as a term for barbarian. In the derogatory interpretation of Vasari, Gothic art was the barbaric art (of the Goths) that had deleted and forgotten the "good" art of the Ancients (Greeks and Romans) until, in the 15th century, the ancient art was reborn (from here the term Renaissance).

Some fractal components are present in the Gothic churches, an example is the *Basilica di San Marco* (*St Mark's Basilica*, 1064-1093), Venice, Italy). The self-similarity is also present inside the Gothic Cathedral. In the Italian Gothic style, some examples show the presence of the fractal components. In Venice, many palaces (Ca' Foscari, Ca' d'Oro, Palazzo Ducale (Doge's Palace), and Giustinian Palace) have a rising fractal structure. Figure 22 shows an example of fractality present in the Venetian palaces.

The Baroque (1600-1750) was born in Italy and adopted in Germany, Netherlands, France, and Spain. Baroque is believed to have been derived from the Italian word "Barocco." It was a word used by the philosophers during the Middle Ages to describe a hindrance in schematic logic. After, this term was used to describe any contorted process of thought or complex idea. Another possible meaning derives from the Spanish "barrueco," the Portuguese form "barroco," used to describe an imperfect or irregularly shaped pearl. This word has survived in the jeweler's term "Baroque pearl." For other sources, the word "baroque" could be derived from the Arabic word "*Burga,*" meaning "uneven surface."

Baroque was also associated with Catholic art, but it progressed and diffused its style into Protestant countries over the centuries. Baroque is a style that expressed power and rigor, "the style of absolutism." It suggested movement in static works of art and impacted significant architectural challenges. Baroque architecture was based on mathematics.

In the age of the Baroque, the architects and the patrons thought of the buildings as "studies in practical mathematics" (this is a phrase of Italian religious Virgilio Spada (1596-1662), that has realized the plan of the *Chapel Spada* located in Rome). The Baroque architecture could be analyzed using a fractal point of view. We can find self-similar shapes; for example, the self-similarity is present in the plans of some churches. *Karlskirche* (*St. Charles's Church*) (1715-1737, Vienna, Austria) is adorned with an oval repeated in three different scales, for instance (Sala, 2005).

The Swiss architect Francesco Borromini (1599-1667) used octagons, Greek crosses, and other shapes for the coffering of the dome in his Roman Catholic church *San Carlo alle Quattro Fontane* (*Saint Charles at the Four Fountains*), also known as *San Carlino* (c. 1634-1644, Rome, Italy).

There is an interesting comparison between the fractal set called "Koch Island" and the Michelangelo project for *St. Peter Basilica.* An analogy of shape is in the plan of *Sant'Ivo alla Sapienza* (1642-1660), a Roman Catholic church in Rome designed by Borromini. Figure 23 shows these analogies (*St. Peter*'s plan is the original).

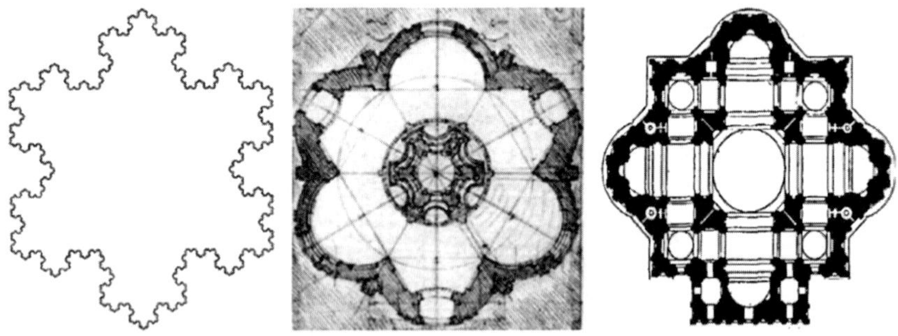

(a) (b) (c)

Figure 23. Koch's Island (a). Borromini's plan of *Sant'Ivo alla Sapienza* (Rome) (b). Michelangelo's project for *St. Peter* (Rome) (c).

Contemporary architecture uses fractal objects as a conscious act of design, thanks to the help of computers, CAD/CAE systems, and new building materials. For example, Polish-American architect Daniel Libeskind has been fascinated by the fractal geometry and the self-similarity. He uses them both in the *Art Museum* (Denver) and the *Victoria & Albert Museum* (London) to create light effects and to highlight the symmetry breaking. Rem Koolhaas, Zaha Hadid (1950-2016), Coop Himmelb(l)au, Jean Nouvel, Renzo Piano, Santiago Calatrava, Zvi Hecker, Steven Holl, Herzog & de Meuron, Massimiliano Fuksas, Tadao Andō, Norman Foster and other "starchitects," find new shapes for their dynamic architectures in the fractal geometry.

Polish-born Israeli architect Zvi Hecker is a poet of form. His works have been recognized for their emphasis on geometry and modularity, but with an increasing asymmetry. Some of his major projects include the *Spiral Apartment House* (1984-1990), a Ramat-Gan (Israel), *Heinz-Galinski School* (1991-1995) in Berlin, the *Holocaust Memorial* (1997) in Berlin, the *City Center* (1996) in Bucharest, the Jewish Culture Center (1996-1999) in Duisburg (Germany). Hecker applied fractal geometry as a common element in his projects, such as breaking symmetry or creating effects that resemble self-similarity. The *Holocaust Memorial* is organized in self-similar structures as if they were fragments of an explosion. His project for the *Heinz-Galinski School* contains a triple fractality. The first one is of a geometric type. In analogy, we can find a fractal shape in the *Nude Descending a Staircase, No. 2*, by Marcel Duchamp. The second is a fractality inspired by nature. The school reproduces a sunflower, representing a symbolic "gift," that the architect makes to the children of Berlin. The third is a "logic" self-similarity: the school is a metaphor for a small city, modeled like a "Russian doll." Bifurcation processes, described through specific fractal systems defined as L-Systems, can be applied in architecture. Figures 24a and 24b illustrate a fractal tree and an example of a tensile structure.

Swiss architect Mario Botta uses forms that can be interpreted using fractal geometry and complexity. The emphasized features in these works are self-similarity, the repetition of the same shape on different scales, and the complexity of the texture. For example, the *MART Museum of Modern & Contemporary Art* (1996–2002) in Rovereto, Italy, is a Botta work that can be analyzed using a fractal perspective (Sala & Cappellato, 2019). The volume of the *MART* is characterized by a large round square with light filtering from above, which is a strong architectural sign, while the exhibition spaces are more anonymous. The impressive mass of the museum is not revealed at first sight but remains partially hidden. Botta found inspiration in classical models (especially the large dome at the Pantheon) and paid special attention to new technical solutions. The radial dome is made of steel and plexiglass and has a missing 'slice,' made possible by complex engineering solutions. Underneath the dome is a square, and in the center, there is a fountain. The complexity is present in the textures of *Garnet Chapel* (Penkenjoch, Zillertal, Austria, 2011-2013). The small chapel is located in the Austrian Alps at 2087 meters above sea level on Mount Penkenjoch, standing out for its very marked and singular geometry. It is located a short distance from the ski resorts present in the area. The project consists of two volumes: a parallelepiped base in reinforced concrete, which supports the main volume shaped as a dodecahedron, which refers to the crystals of "Almandine Garnet" stone, a mineral found in the area. In the base volume, some stairs allow access to the chapel's interior, which is made of a wooden structure covered on the surface with strips of larch wood, accentuating the scansion of the rhomboidal faces of the votive space. The altar is made of walnut wood with an inlaid garnet stone slab in the center by the artist Markus Thurner. The exterior is covered with Corten steel sheets, which can self-protect from corrosion by creating a natural protective layer; natural light falls from a small zenithal oculus located in the center of the rhombus, which incisively marks the surrounding space.

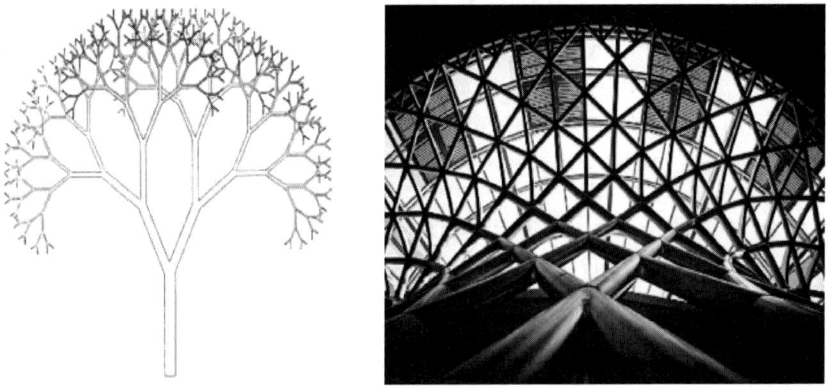

a) b)

Figure 24. A fractal tree (a). Tensile structure (b).

Oceania is the youngest continent, so its architectural history is very recent. In any case, we can find some fractal components in the interior wooden decorations of the *Marae*. It is a sacred building used for ceremonies and meetings. More recently, Danish architect Jørn Oberg Utzon (1918-2008) designed the *Opera House* (1957-1973) in Sydney, Australia, using unconscious fractal geometry through self-resembling forms that recall the valves of a shell, as shown in Figure 25. The Opera House was created from the combination of three distinct parts: the plaza, the auditorium, and the covers (which needed to be different sections of the same ideal sphere corresponding to a diameter of 75 meters).

Figure 25. *Sydney Opera House* (1957-1973, Sydney, Australia) by Utzon. Its shape recalls the valves of a shell.

5. The Box-Counting Dimension

The box-counting dimension is connected to the problem of determining the fractal dimension of a complex two-dimensional image. It is defined as the exponent D_b in the relationship:

$$N(D) \approx \frac{1}{d^{D_b}} \tag{2}$$

N(d) is the number of boxes of linear size d necessary to cover a data set of points distributed in a two-dimensional plane. The basis of this method is that, for objects that are Euclidean, equation (2) defines their dimension. One needs several boxes proportional to 1/d to cover a set of points lying on a smooth line, proportional to $1/d^2$ to cover a set of points evenly distributed on a plane, and so on. Applying the logarithms to equation (2), we obtain N(d) ≈

$-D_b$ log(d). The box-counting dimension can be produced using this iterative procedure:

- superimpose a grid of square boxes over the image (the grid size as given as s_1);
- count the number of boxes that contain some of the image ($N(s_1)$);
- repeat this procedure, changing (s_1), to smaller grid size (s_2);
- count the resulting number of boxes that contain the image ($N(s_2)$);
- repeat this procedure by changing s to smaller and smaller grid sizes.

The box-counting dimension is defined by:

$$D_b = \frac{[log(N(s_2)) - log(N(s_1))]}{\left[log\left(N\left(\frac{1}{s_2}\right)\right) - log\left(N\left(\frac{1}{s_1}\right)\right)\right]} \quad (3)$$

where 1/s is the number of boxes across the bottom of the grid.

We can apply the box-counting dimension in architecture, too. It is calculated by counting the boxes containing lines inside the drawing. The box count was used to elevate Frank Llyod Wright's building (Robie House, 1909) (Bovill, 1996, p. 122).

To determine the degree of complexity in the Mesoamerican arts and architecture, Burkle-Elizondo et al., have collected more than a hundred images of Mesoamerican artistic and architectural works by reviewing the literature on archeology (Burkle-Elizondo & Cepeda-Valdéz, 2002; Burkle-Elizondo et al., 2003; Burkle-Elizondo et al., 2004). All these images have been digitized using a Printer-Copier-Scanner (Hewlett Packard®, Model LaserJet 1100A) and saved in bitmap (*.bmp) format on a Personal Computer (Hewlett Packard®, Model Pavilion 6651). After that, these images were analyzed with the program Benoit®, version 1.3, to calculate Box (D_b), Information (D_i), Mass dimensions (D_M), and their respective standard errors and intercepts on log-log plots. It was considered that the information dimension differs from the box dimension because its boxes contain more points. For all the cases, the fractal dimension values were high, from a Db = 1.8030.023 for the left and superior side of the "Vase of Seven Gods" to a D_M = 2.492 ± 0.195 for the left side of the "Door to Underworld of the Temple 11, platform" at Copán.

The degree of complexity found in the Mesoamerican buildings can be explained if we remember the basis of Mesoamerican Cosmo vision and the idea of how the Universe works, which have influenced this architecture (Burkle-Elizondo & Cepeda-Valdéz, 2002; Burkle-Elizondo et al., 2003; Burkle-Elizondo et al., 2004; Burkle-Elizondo et al., 2015; Burkle-Elizondo & Sala, 2004).

Conclusion

In this chapter, we have outlined an approach that focuses on analyzing the complexity and fractality from three perspectives:

1. Exploring complexity and fractality in the shapes of buildings.
2. Identifying complexity and fractality within the structures of buildings.
3. Assess the level of complexity in a building using the box-counting dimension.

Will complexity and fractality impact architecture? We agree with Alexander that the 20th century was dominated by physics, while the 21st century will be defined by biology. The shift toward complexity indicates that new techniques, perspectives, modes of thinking, and approaches to action are necessary (Alexander, 2002-2005). We are witnessing the beginning of this transformation.

References

Alexander, C. (2002-2005). *The Nature of Order* (4 volumes), Berkeley: Center for Environmental Structure (CES).

Birkhoff, G. D. (1933). *Aesthetic Measure*. Cambridge, MA: Harvard University Press.

Blackwell, W. (1984). *Geometry in Architecture*, London: John Wiley & Sons.

Bovill, C. (1996). *Fractal Geometry in Architecture and Design*, Boston: Birkhäuser.

Burkle-Elizondo, G. & Cepeda-Valdéz R. (2002). Do The Artistic and Architectural Works Have Fractal Dimension? *Emergent Nature*, M. M. Novak (ed.), Singapore: World Scientific, pp. 431-432.

Burkle-Elizondo, G. & Sala, N. (2004). Complexity and Fractal Dimension in 26 Mesoamerican Pyramids, *Proceedings 7th Generative Art Conference,* Milano, Italy, vol. 1, pp. 206-214

Burkle-Elizondo, R., Sala, N. & Cepeda-Valdéz, R. (2003). Complexity in the Mesoamerican Artistic and Architectural Works, *Programme and Abstracts*

International Nonlinear Sciences Conference (Research and Applications in the Life Sciences), Vienna, Austria, p. 18.

Burkle-Elizondo, R., Sala, N. & Cepeda-Valdéz, R. (2004). Geometric And Complex Analyses Of Maya Architecture: Some Examples. *Nexus V: Architecture and Mathematics*, ed. K. Williams K., Fucecchio, Italy: Edizioni Dell'Erba, pp. 57-68.

Burkle-Elizondo, G., Sala, N., & Valdez-Cepeda, R. D. (2015). Geometric and Complex Analyses of Maya Architecture: Some Examples. In: Williams, K., Ostwald, M. (eds) *Architecture and Mathematics from Antiquity to the Future*. Basel: Birkhäuser, pp. 113-125.

Casti, J. L. "Complexity." *Encyclopedia Britannica*, 23 August 2023, https://www.britannica.com/science/complexity-scientific-theory. Accessed 10 March 2024.

Dunlap, R. A. (1998). *The Golden Ratio and Fibonacci Numbers*, Singapore: World Scientific.

Eaton, L. K. (1998). Fractal Geometry in the Late Work of Frank Llyod Wright: the Palmer House. *Nexus II: Architecture and Mathematics*, ed. K. Williams, Fucecchio,Italy: Edizioni Dell'Erba, pp. 23 - 38.

Eglash, R. (1999). *African fractals: Modern computing and indigenous design.* New Brunswick: Rutgers University Press.

Ghyka, M. (1977). *The Geometry of Art and Life*, New York: Dover Publications.

Hargittai, I. & Hargittai, M. (1996). *Symmetry: A Unifying Concept*, New York: Random House.

Huberman, B. A. & Hogg, T. (1986). Complexity and adaptation. *Physica* D 22, 376–384.

Jackson, W. J. (2002). *Hindu Temple Fractals*. Retrieved from: https://www.academia.edu/347639/Hindu_Temple_Fractals.

Jencks, C. (1998). Complexity definition and nature's complexity, *Architectural Design*, n. 129, pp. 8–10.

Portoghesi, P. (1999). *Natura e Architettura*, Milano: Skira, (English version: *Nature and Architecture*, Skira, Milano, 2000).

Sala, N. & Cappellato, G. (2003). *Viaggio matematico nell'arte e nell'architettura*, Milano: Franco Angeli.

Sala, N. & Cappellato, G. (2004). *Architetture della complessità,* Milano: Franco Angeli.

Sala, N. & Cappellato. G. (eds.) (2018). *Chaos and Complexity In the Arts and in Architecture: Research in Progress*. New York: Nova Publishers.

Sala, N. & Cappellato, G. (2019). *Connections between Mathematics, the Arts, and Architecture,* New York: Nova Science Publishers.

Sala, N. (2002). The presence of the Self- Similarity in Architecture: Some examples. M. M. Novak (ed.), *Emergent Nature*, Singapore: World Scientific, pp. 273 – 283.

Sala, N. (2003). Fractal geometry and self-similarity in architecture: an overview across the centuries. In Meeting Alhambra, *ISAMA-BRIDGES Conference Proceedings*. pp. 235–244.

Sala, N. (2005). Fractal Components in the Gothic and the Baroque Architecture *Proceedings 8h Generative Art Conference,* Milano, Italy, vol. 1, pp. 297–315.

Sala, N. (2006). Fractal Geometry and Architecture: Some Interesting Connections, *Eco Architecture 2006*, Southampton: WIT Press, pp. 163–174.

Sala, N. (ed.) (2007). *Chaos and Complexity in Arts and Architecture*, New York: Nova Science Publishers.

Sala, N. (2012). Complexity and Architecture, *Chaos and Complexity Letters,* vol. 6, issue 1-2, pp 93–102.

Sala, N. (2018). Complexity and Architecture. *Chaos and Complexity in the Arts and Architecture: Research in Progress*, New York: Nova Science Publishers, pp. 117–128.

Sporns, O. (2007). "Complexity." *Scholarpedia*, 2(10):1623. http://www.scholarpedia.org/article/Complexity. Accessed 10 January 2024.

Venturi, R. (1992). *Complexity and contradiction in architecture*. New York: The Museum of Modern Art.

Chapter 7

A New Integrated Model of Health Derived from Telomere Science: The “Chain” Model

Massimo Agnoletti*, PhD
Verona University, Verona, Italy

Abstract

The telomere “funnel effect” describes the convergence, partial independence, and equivalence of nine main factors (nutritional, physical activity, psychological, etc.) on telomeres, the chromosomal structures representing our biological clock, because they determine our longevity and overall quality of life.

The telomere “funnel effect” has as its logical consequence a specific bio-psycho-social model characterized by the equivalence of the nine main factors (genetic and epigenetic) that influence our health and well-being.

The “chain” model derived from this telomeres’ “funnel effect” precisely describes the partial independence and equivalence of the nine main factors acting in our organism.

The characteristics of the “chain” model of human health and well-being provide for specific dynamics such as, for example, the fact that “the chain is only as strong as the weakest of its links,” so it only takes the scarcity of one factor (nutritional, physical activity, psychological, etc.) to accelerate the aging of our organism resulting in reduced longevity and increased likelihood of developing a range of diseases.

This a-hierarchical, highly interconnected, and integrated model of human well-being and health has relevant clinical implications, partly contradicting some widely held assumptions even within the health

* Corresponding Author’s Email: max.agnoletti@gmail.com

In: Complexity and Fractality in Human Change and Beyond
Editor: Nicoletta Sala
ISBN: 979-8-89530-575-1

professional community. This work describes some of these very impactful implications on the actual strategies envisioned by this original model to improve our human well-being and health.

Keywords: telomere, complexity, health model, chain model of human health

1. Introduction

The scientific literature has identified the fundamental role of telomeres in protecting chromosomes by ensuring their genetic stability and influencing human health and well-being (Armanios & Blackburn, 2012; Aubert & Lansdorp, 2008; Blackburn, Epel & Lin, 2015; Cawthon et al., 2003; Fitzpatrick et al., 2007; Valdes et al., 2005).

These biological structures, located at the ends of each chromosome, are composed, in the human species, of repeated noncoding nucleobase sequences (5'-TTAGGG-3') that ensure the structural solidity of the rest of the chromosome.

With each cellular replication, telomeres shorten reaching a maximum number of replications (the so-called "Hayflick limit"), which corresponds to a certain length of the telomeres themselves (and thus a certain number of nucleobases), beyond which sufficient structural maintenance capacity of the chromosome is no longer guaranteed (Hayflick & Moorhead, 1961). In other words, if initially, the zygote cell has a telomere length of about 20 thousand pairs of DNA bases, already at the time of child delivery these structures are composed of about 15 thousand nucleobases but the moment this amount reaches about 5 thousand bases it happens that the telomere structure is no longer sufficient to guarantee the structure of the chromosome.

2. Telomeres

To help people understand the function and structure of telomeres, the metaphor of the plastic tips of shoelaces is often used. When these tips are intact and in good condition, they ensure that the entire lace is appropriately used. However, if they deteriorate beyond a certain point, the lace begins to fray progressively.

In this metaphor, the lace represents our chromosomes, and their fraying depicts the progressive aging process, the structural instability of the chromosome with the predisposition to disease development and subsequent cell death.

Telomere length is crucial in determining cellular longevity and the overall health of the whole organism for this reason telomere length is currently considered our most reliable biological "clock." Telomere length is influenced by the activity of the enzyme telomerase, because the role of this biological structure is to maintain and rebuild, by adding nucleobases bases, telomeres thereby counteracting their physiological shortening (Andrews & Cornell, 2014; Andrews & Cornell, 2017; Armanios, 2013; Blackburn, 1991; Blackburn, 2010).

We already know from the solid scientific literature that several factors such as healthy lifestyles significantly affect telomerase activity, and thus telomere length, influencing longevity, health, and susceptibility to aging-related diseases (Cherkas et al., 2008; D'Mello et al., 2015; Shalev et al., 2013; Puterman et al., 2010; Tucker, 2017; Valdes et al., 2005).

Interestingly, while it is true that we are born with a certain telomere length and that it gradually shortens over a person's life span, it is also true that the relative speed of the "consumption" of telomeres is extremely variable and dependent on various factors that affect telomerase activity.

While some factors "accelerate" telomere shortening because they limit the maintenance action by telomerase (for example smoking, little sleep, sedentariness, chronic psychosocial distress, poor diet, etc.) other factors "slow" its consumption because they more effectively promote telomerase activity (meditation, regular physical activity, proper nutrition, proper sleep quality, etc.) (Cherkas et al., 2008; Epel et al., 2016; Jackowska et al., 2012; Valdes et al., 2005). This dynamic concerning the accelerating and slowing action in telomere consumption was called by Dr. Andrew Bill (discoverer of telomerase) the "tug-of-war" of telomeres because it describes the opposing action of different factors that determine telomere length (Andrews & Cornell, 2017). Figure 1 shows the scheme of telomeres and telomerase.

In this perspective we could say that the general therapeutic goal, prevention, and health promotion consists of promoting in people the physiological telomere shortening in the slowest possible way by supporting the maintenance capacity of telomerase to prolong as much as possible the residual longevity and consequently the quality of life in terms also of reducing the probability of developing cardiocirculatory, immune problems, etc. (Agnoletti, 2018a; Agnoletti, 2018b; Agnoletti, 2018c; Agnoletti, 2018d).

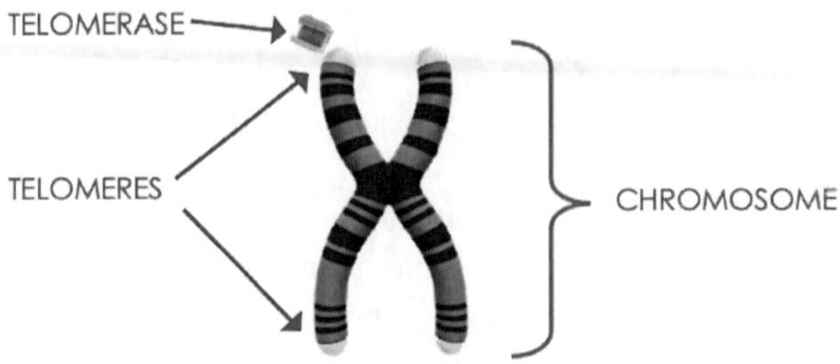

Figure 1. Telomeres and telomerases. Telomeres protect chromosomes, and telomerase maintains and rebuilds by adding nucleobase bases. Telomeres counteract their physiological shortening.

We have seen that the initial length of telomeres is an unmodifiable parameter however the rate of its consumption is strongly influenced by exposure to factors such as psychological distress, environmental pollutants, smoking or alcohol abuse, and many other elements that help define the rate at which our telomeres shorten ultimately determining our longevity (Armanios & Blackburn, 2012; Blackburn, Epel & Lin, 2015; Calado & Young, 2009; López-Otín et al., 2013).

Analyzing the scientific literature, I identified nine main factors that result in significant changes in telomere length:

1. genetics.
2. psychological well-being.
3. nutrition.
4. social support.
5. circadian rhythms.
6. the physic-chemical environment.
7. the microbiota.
8. sleep.
9. physical activity.

Indeed, the scientific literature has shown that the genetic factor (Armanios & Blackburn, 2012; Calado & Young, 2009; Harley, Futcher & Greider, 1990; Liu & Khaing, 2012), psychological well-being (Epel et al., 2004; Epel et al., 2016; Gotlib & LeMoult, 2019; Price et al, 2013; Ridout et al., 2016; Zhang et al., 2016), the way we feed ourselves (Crous-Bou et al., 2014; Garcia-Calzon et al., 2014; Leung et al., 2018; Marcon et al.,2012; Niles & Terry, 2016), the quality of the social network we perceive (Needham et al.,

2015). O'Donovan et al., 2013; Puterman et al., 2013; Shalev et al., 2013; Steptoe et al., 2004), circadian rhythms (Carroll et al., 2016; Savolainen et al., 2014; Simon et al., 2006; Wen et al., 2015; Zhu et al, 2019), the physicochemical quality of the environment in which we live (Blackburn & Epel, 2015; Clement et al., 2020; Dioni et al., 2011; Hoxha et al., 2009; McGrath, Jones & Sorkin, 2017; Ward-Caviness et al, 2016), the composition of the microbiota that colonizes our bodies (Ghosh et al., 2017; Liu et al., 2020; Jiang et al., 2017; Vaiserman & Krasnienkov, 2021), sleep quality (Al-Hazzouri, et al., 2014; Carroll et al., 2016; Jackowska et al., 2012; Li & Wang, 2020; Prather et al, 2011) and the physical activity that we do (Cherkas et al., 2008; Lynch et al., 2020; Munstock et al., 2021; Puterman et al., 2010; Tejada-Garrido, Martín-Vallejo, & Navarro-López, 2024) they all contribute to the rate of telomere consumption and thus life expectancy (in quantitative and qualitative terms).

Of course, all these factors, in part, influence each other (for example, physical activity influences sleep quality or circadian rhythms influence the metabolic impact of diet) however, each of these factors also has a certain autonomy in terms of their complex function and characteristic causal mechanism influencing telomeres.

For example, psychological stress management and physicochemical environmental quality are both factors that influence telomere length. However, they do so partly independently of each other because they follow at least partly different and autonomous causal pathways.

For this chapter, it is particularly interesting to note that currently the scientific literature does not highlight the dominance of one factor more important than the others but describes a dose-dependent quantitative effect for each (Agnoletti, 2019; Blackburn, Epel & Lin, 2015; Bojesen, 2013; Epel et al., 2018; Lin, Epel, & Blackburn, 2012; Liu et al., 2018; Shammas, 2011; Starr et al., 2008; Valdes et al., 2005).

Because of this lack of dominance of one factor over the others (what we might call the "equivalence principle" of the factors considered), the phenomenon of accelerating or slowing down telomere consumption appears to be, at least in part, independent of the type of specific mechanism that leads to altering telomerase function and thus telomere length.

Consequently, the phenomenon of accelerating or slowing telomere aging appears to be at least partly independent of the type of "upstream" informational pathway that impacts telomerase activity.

3. The Funnel Effect of Telomeres

In this regard, I coined the term "funnel effect" or "bottleneck effect" of telomeres (Agnoletti, 2019) to describe the equivalent and convergent nature of all nine macro-factors identified in the scientific literature (physical activity, nutrition, psychological well-being, sleep quality, etc.) that, at least in part, independently influence telomere length.

The implications of the funnel effect are many and involve both strictly clinical and nonclinical aspects (Agnoletti, 2021; Agnoletti, 2022).

One of the most important implications of the "funnel effect" is that, in the context of the bio-psycho-social paradigm (Engels, 1977; Schwartz, 1982), it defines a specific model of human well-being and health that can be called a "chain" model.

Since there is no one factor dominant over the others, but all factors contribute equivalently to influence telomere length ("equivalence principle"), the "chain" model predicts that human health can be represented by a chain consisting of nine links where each link in the chain represents a specific factor (psychological well-being, physical activity, nutrition, sleep quality, etc.).

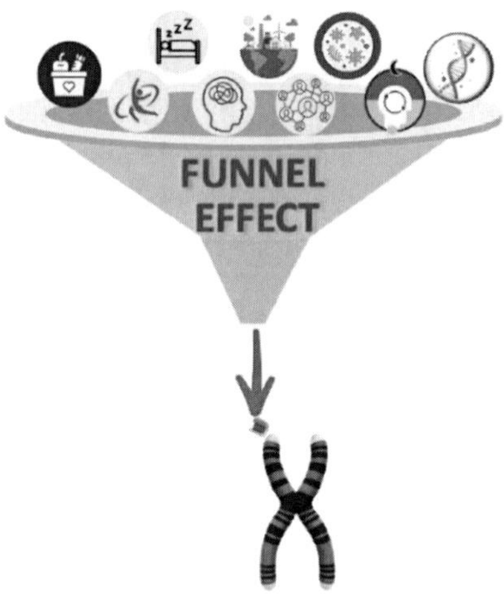

Figure 2. Funnel effect of telomeres. The funnel effect of telomeres describes the equivalent and convergent nature of all nine macro-factors that, at least in part, independently influence telomere length.

This specific model, which reflects the evidence described above from telomere science, is consistent with the integrated approach envisioned by the bio-psycho-social paradigm and has aspects that partially contrast the hierarchical models and hyper-specialized culture related to human health and well-being.

In the current historical period, hierarchical models and hyper-specialist culture are widely shared even within clinical professional communities (at least in the Western world).

Figure 2 illustrates the funnel effect of telomeres.

4. The Chain Model

Let us, therefore, look at some relevant clinical implications of the bio-psycho-social "chain" model of human well-being and health. The "chain" model proposed here implies that:

1. an overall assessment of the nine factors described is needed initially.
2. It is necessary to prioritize intervention to the factors that are most scarce (because they represent the factors or factors that represent the greatest vulnerability for telomeres).
3. The overall therapeutic goal should not be "optimal" attainment in one, or more factors, but avoidance of "scarcity" in one, or more factors.

If the chain metaphor captures the interconnected and partially independent nature of the nine main factors that influence telomere length, it follows that the priority for assessing the overall state of our health is to evaluate all nine of these factors simultaneously in order to identify the weakest link; or links, because the logical architecture of this model implies that "the chain is only as strong as the weakest of its links."

Wellness and human health professionals should have as an initial priority the comprehensive and integrated assessment envisioned by the "chain" model just described, to identify the factor, if any, that represents the greatest vulnerability to telomeres for the patient. Of course, this priority should be considered regardless of each practitioner's specialized field. Figure 3 shows the chain model of human health.

Indeed, if what is described by the "funnel effect" and predicted by the "chain" model of human health is true, then it makes little sense to intervene by focusing solely on, for example, the psychological level, or exclusively on sleep quality, or physical, or nutritional since the impact on telomeres follows the logic described by the nine-factor "equivalence" principle.

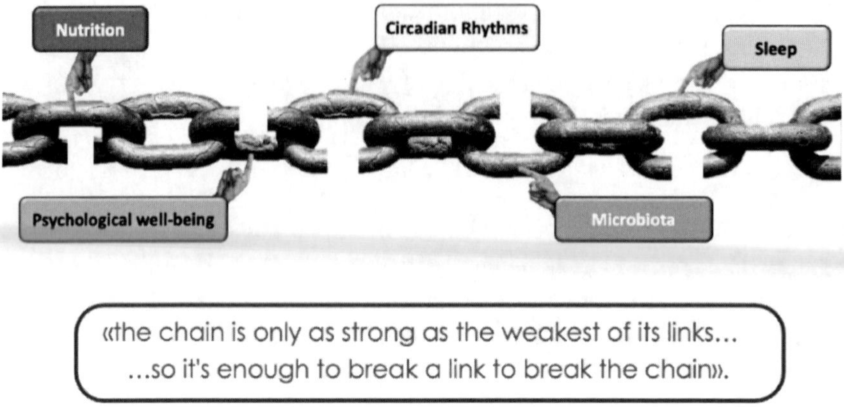

Figure 3. The Chain model of human health. The logical architecture of the "chain" model implies that the "chain" of our health consists of nine links; each link corresponds to one of the nine macro factors.

For the practitioner, focusing exclusively on one's specific area of intervention (e.g., psychological or nutritional) without taking into account the dynamics described by the "chain" model, risks being ineffective in the medium to long term because by underestimating one or more "links" (e.g., that related to sleep quality or circadian rhythms) that would instead require more timely intervention, one does not counteract the factor, or factors, that contribute most to shortening telomeres.

Instead, the intervention of professionals should tend to "strengthen" the weakest link(s) to achieve at least a degree of sufficiency (for example, a minimum quality of sleep or physical activity) but not aim to "excel" in one or more factors at the expense of others (for example when a person significantly increases his or her physical activity but simultaneously begins to eat incorrectly). Not to make an initial comprehensive analysis of the main nine factors envisaged by the "chain" model but to continue to apply a myopic hyper-specialized view of human health thus also implies a potential risk represented by the fact that the time and energy invested in a "link" that does not possess the correct priority of intervention is actually time and energy taken away from the weaker "link" that impacts, at that specific moment in life, more on telomere shortening and thus on the person's life expectancy.

In the metaphor of the "chain" model, we could say that the practitioner's goal is to prevent there from being "paper" links rather than reinforcing one or more links to make them "titanium."

In the chain's logic, a ring, or rings, "made of titanium" coupled with one or more "paper" rings still makes the chain very fragile.

Because of the overall, integrated nature of the chain ("the chain is only as strong as the weakest of its links"), the best strategy that optimizes its strength is to strengthen the weakest links until they become as strong as the strongest links.

A chain composed of nine iron links is much stronger than a chain composed of eight titanium links and one paper link.

The characteristic dynamics of the "chain" model also seem to be more realistic in reflecting the complexity and heterogeneity of human behavior and performance; in fact, it turns out to be more explanatory than hierarchical models involving the dominance of one factor over the others. For example, the cases at the antipodes of the premature deaths of professional sportsmen who conducted intense physical activity for many years and followed proper nutrition and the cases of ultra-centenarians who, despite never having engaged in particularly intense physical activity (but never having been sedentary) and despite suboptimal nutrition (but never having been poor for long periods) represent performances of longevity to the very distant that can be understood only by the "chain" model and not by the hierarchical models that instead involve, for example, diet or physical activity as the dominant factors with more impact (than other factors) on human health and well being.

Conclusion

The "chain" model, here described, clearly represents a cultural challenge for the current hyper-specialized and ultra-sectoral education that has been widely shared by both undergraduate and postgraduate professional communities for decades.

Current hyper-specialist paradigms are declined into clinical protocols used by professionals who, as a result, tend to underestimate significantly the integrated and global dimension of people in supporting their patients. Although over the past century, the biomedical advances recorded in dealing with acute and traumatic issues are obvious and worthy of admiration and gratitude, the equally obvious low effectiveness of interventions in managing current chronic issues reflects this reductionist, specialized, and hierarchical approach in some respects opposite to the one described here as the "chain" model.

Should the "chain" model prove to be valuable from a clinical point of view, it would be desirable for it to be integrated as soon as possible into the educational pathways of health and human welfare professionals but also promoted and shared through appropriate communications aimed at citizens.

References

Agnoletti, M. (2022). L'effetto imbuto dei telomeri sui protocolli di supporto psicofisico. *State of Mind*, 5. https://www.stateofmind.it/2022/05/telomeri-effetto-imbuto/.

Agnoletti, M. (2021). L'olobionte umano-microbiota e l'effetto imbuto dei telomeri. *State of Mind*, 10. https://www.stateofmind.it/2021/10/olobionte-microbiota-telomeri/.

Agnoletti, M. (2019). L'Effetto Imbuto o Effetto Collo di Bottiglia dei telomeri. *Medicalive Magazine*, 5, 18-21. https://www.medicalive.it/leffetto-imbuto-o-effetto-collo-di-bottiglia-dei-telomeri/.

Agnoletti, M. (2018a). L'Asse psiche-telomeri ecco come la mente influenza l'invecchiamento. *PNEINEWS*, 5, 4-8.

Agnoletti, M. (2018b). La Scienza dei Telomeri. Come sta cambiando radicalmente il concetto d'invecchiamento con importanti conseguenze sul piano della salute e del benessere psicofisico umano. *Medicalive Magazine*, 8, 5-7. https://www.medicalive.it/scienza-telomeri-agnoletti/.

Agnoletti, M. (2018c). La nuova frontiera della psicologia: la Psicologia Epigenetica. *State of Mind*, 10. https://www.stateofmind.it/2018/10/psicologia-epigenetica.

Agnoletti, M. (2018d). La Scienza dei Telomeri e il modello integrato di salute psicofisica. *Medicalive Magazine*, 11, 8-10. https://www.medicalive.it/la-scienza-dei-telomeri-e-il-modello-integrato-di-salute-psicofisica/.

Al-Hazzouri, A. Z., Stone, K. L., Hough, C. M., Yaffe, K., & Stewart, A. L. (2014). Sleep duration and telomere length in a sample of older men: results from the Osteoporotic Fractures in Men Study. *Journal of Aging Research*, 2014, 819341.

Andrews, B. & Cornell, J. (2014). *Telomere Basics: Curing Aging*. Nevada, USA: Sierra Science.

Andrews, B. & Cornell, J. (2017). *Telomere Lengthening*. Nevada, USA: Sierra Science.

Armanios, M., & Blackburn, E. H. (2012). The telomere syndromes. *Nature Reviews Genetics*, 13(10), 693–704.

Aubert, G., & Lansdorp, P. M.(2008). Telomeres and aging. *Physiological Reviews*, 88(2), 557-579.

Blackburn, E. H., & Epel, E. S. (2015). Telomeres and adversity: Too toxic to ignore. *Cell*, 162(4), 709–711.

Blackburn, E. H., Epel, E. S., & Lin, J. (2015). Human telomere biology: A contributory and interactive factor in aging, disease risks, and protection. *Science*, 350(6265), 1193–1198.

Bojesen, S. E. (2013). Telomeres and human health. *Journal of Internal Medicine*, 274(5), 399-413.

Cacioppo, J. T., Hawkley, L. C., & Thisted, R. A. (2010). Perceived social isolation makes me sad: five year cross-lagged analyses of loneliness and depressive symptomatology in the Chicago Health, Aging, and Social Relations Study. *Psychology and Aging*, 25(2), 453–463.

Calado, R. T., & Young, N. S. (2009). Telomere diseases. *New England Journal of Medicine*, 361(24), 2353-2365.

Carroll, J. E., Esquivel, S., Goldberg, A., Seeman, T. E., Effros, R. B., Dock, J., ... & Olmstead, R. (2016). Insomnia and telomere length in older adults. *Sleep*, 39(3), 559–564.

Carroll, J. E., Seeman, T. E., Olmstead, R., Melendez, G., Sadakane, R., & Freedland, K. E. (2016). Improved sleep, psychological well-being, and circadian rhythms predict long-term improvements in telomere length in caregivers of patients with Alzheimer's disease. *Sleep*, 39(9), 1503–1511.

Cawthon, R. M., Smith, K. R., O'Brien, E., Sivatchenko, A., & Kerber, R. A. (2003). Association between telomere length in blood and mortality in people aged 60 years or older. *The Lancet*, 361(9355), 393–395.

Cherkas, L. F., Hunkin, J. L., Kato, B. S., Richards, J. B., Gardner, J. P., Surdulescu, G. L., ... & Aviv, A. (2008). The association between physical activity in leisure time and leukocyte telomere length. *Archives of Internal Medicine*, 168(2), 154–158. https://doi.org/10.1001/archinternmed.2007.39.

Clemente, D. B. P., Vrijheid, M., Martens, D. S., Plusquin, M., Nawrot, T. S., & Janssen, B. G. (2020). Prenatal ambient air pollution, placental mitochondrial DNA content and telomere length: An ENVIRONAGE birth cohort study. *Environment International,* 138, 105597.

Crous-Bou, M., Fung, T. T., Prescott, J., Julin, B., Du, M., Sun, Q., ... & De Vivo, I. (2014). Mediterranean diet and telomere length in Nurses' Health Study: population-based cohort study. *BMJ,* 349, g6674.

D'Mello, M. J., Ross, S. A., Briel, M., Anand, S. S., Gerstein, H., & Paré, G. (2015). Association between shortened leukocyte telomere length and cardiometabolic outcomes: Systematic review and meta-analysis. *Circulation: Cardiovascular Genetics*, 8(1), 82-90.

Dioni, L., Hoxha, M., Nordio, F., Tarantini, L., Albetti, B., Grillo, G., ... & Baccarelli, A. A. (2011). Effects of short-term exposure to inhalable particulate matter on telomere length, telomerase expression, and telomerase methylation in elderly adults. *Environmental Health Perspectives*, 119(5), 737-742.

Engel, G. L. (1977). The need for a new medical model. A challenge for biomedicine. *Science*. 196, 129–136.

Epel, E. S., & Prather, A. A. (2018). Stress, telomeres, and psychopathology: Toward a deeper understanding of a triad of early aging. *Annual Review of Clinical Psychology*, 14, 371–397.

Epel, E. S., Blackburn, E. H., Lin, J., Dhabhar, F. S., Adler, N. E., Morrow, J. D., & Cawthon, R. M. (2004). Accelerated telomere shortening in response to life stress. *Proceedings of the National Academy of Sciences*, 101(49), 17312–17315.

Epel, E. S., Puterman, E., Lin, J., Blackburn, E. H., Lum, P. Y., Beckmann, N. D., ... & Mangino, M. (2016). Meditation and vacation effects have an impact on disease-associated molecular phenotypes. *Translational Psychiatry*, 6(8), e880.

Fitzpatrick, A. L., Kronmal, R. A., Gardner, J. P., Psaty, B. M., Jenny, N. S., Tracy, R. P., ... & Aviv, A. (2007). Leukocyte telomere length and cardiovascular disease in the Cardiovascular Health Study. *American Journal of Epidemiology*, 165(1), 14–21.

Garcia-Calzon, S., Moleres, A., Marcos, A., Campoy, C., Martinez-Gonzalez, M. A., Azcona-San Julian, M. C., ... & Marti, A. (2014). Telomere length as a biomarker for adiposity changes after a multidisciplinary intervention in adolescents: The EVASYON study. *PLOS ONE*, 9(2), e89828.

Ghosh, S., Sinha, J. K., Raghunath, M., & Chakrabarti, S. (2017). Increase in the gut microbiota diversity in short-lived ERCC1−/− mice fed with a probiotic mix correlates with improved health and lifespan: Prospects of microbe-based interventions to counteract aging. *Ageing Research Reviews*, 36, 62-74.

Gotlib, I. H., & LeMoult, J. (2019). Telomere length and early life stress: An overview of research and the need for integrated approaches. *Journal of Behavioral Medicine*, 42(4), 535-543.

Harley, C. B., Futcher, A. B., & Greider, C. W. (1990). Telomeres shorten during ageing of human fibroblasts. *Nature*, 345(6274), 458-460.

Hayflick, L., & Moorhead, P. S.(1961). The serial cultivation of human diploid cell strains. *Experimental Cell Research*, 25(3), 585–621.

Hoxha, M., Dioni, L., Bonzini, M., Pesatori, A. C., Fustinoni, S., Cavallo, D., ... & Baccarelli, A. (2009). Association between leukocyte telomere shortening and exposure to traffic pollution: A cross-sectional study on traffic officers and indoor office workers. *Environmental Health*, 8, 41.

Jackowska, M., Hamer, M., Carvalho, L. A., & Steptoe, A. (2012). Short sleep duration is associated with shorter telomere length in healthy men: findings from the Whitehall II cohort study. *PLOS ONE*, 7(10), e47292.

Jiang, X., Zhong, W., Zhang, G., Zheng, Z., Lan, Z., & Xu, L. (2017). Association between gut microbiota and telomere length in colorectal cancer patients. *Journal of Digestive Diseases*, 18(12), 580–585.

Leung, C. W., Fung, T. T., McEvoy, C. T., Lin, J., Epel, E. S., & Dieterich, D. M. (2018). Associations of dietary intake and dietary patterns with leukocyte telomere length in a multi-ethnic sample of older adults: the Multi-Ethnic Study of Atherosclerosis (MESA). *American Journal of Clinical Nutrition*, 108(2), 165-175.

Li, J., Hand, G. A., & Wang, X. (2020). The impact of poor sleep on telomere length: A systematic review. *Journal of Clinical Sleep Medicine*, 16(6), 981-987.

Lin, J., Epel, E., & Blackburn, E. (2012). Telomeres and lifestyle factors: Roles in cellular aging. *Mutation Research/Fundamental and Molecular Mechanisms of Mutagenesis*, 730(1-2), 85–89.

Liu, T., Hougen, H. Y., Qi, X., & Yu, J. (2020). Associations between gut microbiota and telomere biology: new insights into the impact of diet and lifestyle. *Nutrients*, 12(6), 1868.

Liu, Y., & Khaing, A. A. (2012). Telomere dysfunction in human diseases: A review of recent progresses in experimental and clinical research. *Journal of Genetics and Genomics*, 39(1), 9-22.

Liu, Y., Cao, H., Song, S., & Miao, Y. (2018). Multiple environmental and genetic factors act in an integrated manner to negatively regulate telomere length and protect telomere capping in Arabidopsis. *Frontiers in Plant Science*, 9, 1444.

Lynch, B. M., et al., (2020). Physical activity and telomere length in early-stage breast cancer survivors. *Breast Cancer Research*, 22, 56. https://doi.org/10.1186/s13058-020-01298-0.

Marcon, F., Siniscalchi, E., Crebelli, R., Veneziano, S., Pensa, A., & Sgura, A. (2012). Diet-related telomere shortening and chromosome stability. *Mutagenesis*, 27(1), 49-57.

McGrath, E. E., Jones, A., & Sorkin, J. D. (2017). Air pollution and telomere length: A systematic review of the literature. *Journal of Public Health Research*, 6(3), 1040.

Munstock, L., et al., (2021). Physical activity on telomere length as a biomarker for aging: A systematic review. *Sports Medicine - Open*, 7(1), 41. https://doi.org/10.1186/s40798-021-00315-3.

Needham, B. L., Mezuk, B., Bareis, N., Lin, J., Blackburn, E. H., & Epel, E. S. (2015). Depression, anxiety and telomere length in young adults: evidence from the National Health and Nutrition Examination Survey. *Molecular Psychiatry*, 20(4), 520–528.

Niles, J. K., Joshi, A. D., & Terry, M. B. (2016). Diet, telomere length, and cancer risk: A review of potential mechanisms and epidemiological evidence. *Cancer Causes & Control*, 27(6), 845–858.

O'Donovan, A., Slavich, G. M., Epel, E. S., & Neylan, T. C. (2013). Exaggerated neurobiological sensitivity to threat as a mechanism linking anxiety with increased risk for diseases of aging. *Neuroscience & Biobehavioral Reviews*, 37(1), 96–108.

Prather, A. A., Puterman, E., Lin, J., O'Donovan, A., Krauss, J., Tomiyama, A. J., ... & Epel, E. S. (2011). Shorter leukocyte telomere length in midlife women with poor sleep quality. *Journal of Aging Research*, 2011, 721390.

Price, L. H., Kao, H. T., Burgers, D. E., Carpenter, L. L., & Tyrka, A. R. (2013). Telomeres and early-life stress: An overview. *Biological Psychiatry*, 73(1), 15-23.

Puterman, E., Epel, E. S., O'Donovan, A., Prather, A. A., Aschbacher, K., Dhabhar, F. S., ... & Blackburn, E. H. (2013). Predictors of telomere length in the Health and Retirement Study: New insights into resilient aging. *Psychoneuroendocrinology*, 38(9), 2418-2427.

Puterman, E., Lin, J., Blackburn, E., O'Donovan, A., Adler, N., & Epel, E. (2010). The power of exercise: buffering the effect of chronic stress on telomere length. *PLoS ONE*, 5(7), e10837. https://doi.org/10.1371/journal.pone.0010837.

Ridout, K. K., Ridout, S. J., Price, L. H., Sen, S., & Tyrka, A. R. (2016). Depression and telomere length: A meta-analysis. *Journal of Affective Disorders*, 191, 237-247.

Savolainen, K., Eriksson, J. G., Kajantie, E., Lahti, J., Lahti, M., Heinonen, K., ... & Räikkönen, K. (2014). Telomere length and hypothalamic-pituitary-adrenal axis response to stress in elderly adults. *Psychoneuroendocrinology*, 47, 12-17.

Schwartz, G. E. (1982). Testing the biopsychosocial model: The ultimate challenge facing behavioral medicine? *Journal of Consulting and Clinical Psychology*, 50(6): 1040-1053.

Shalev, I., Entringer, S., Wadhwa, P. D., Wolkowitz, O. M., Puterman, E., Lin, J., ... & Epel, E. S. (2013). Stress and telomere biology: A lifespan perspective. *Psychoneuroendocrinology*, 38(9), 1835-1842.

Shalev, I., Moffitt, T. E., Sugden, K., Williams, B., Houts, R. M., Danese, A., ... & Caspi, A. (2013). Exposure to violence during childhood is associated with telomere erosion from 5 to 10 years of age: A longitudinal study. *Molecular Psychiatry*, 18(5), 576–581.

Shammas, M. A. (2011). Telomeres, lifestyle, cancer, and aging. *Current Opinion in Clinical Nutrition & Metabolic Care*, 14(1), 28-34.

Simon, N. M., Smoller, J. W., McNamara, K. L., Maser, R. S., Zalta, A. K., Pollack, M. H., ... & Wong, K. K. (2006). Telomere shortening and mood disorders: preliminary support for a chronic stress model of accelerated aging. *Biological Psychiatry*, 60(5), 432–435.

Starr, J. M., Shiels, P. G., Harris, S. E., & Deary, I. J. (2008). Telomere length and lifetime cognitive change: Combined effect of early and later life adversity. *Neurobiology of Aging*, 29(2), 269–276.

Steptoe, A., Owen, N., Kunz-Ebrecht, S. R., & Brydon, L. (2004). Loneliness and neuroendocrine, cardiovascular, and inflammatory stress responses in middle-aged men and women. *Psychoneuroendocrinology*, 29(5), 593-611.

Tejada-Garrido, C. I., Martín-Vallejo, J., & Navarro-López, V. (2024). Effects of physical exercise on telomere length in healthy adults: Systematic review, meta-analysis, and meta-regression. *JMIR Public Health and Surveillance*, 10, e46019. https://doi.org/10.2196/46019.

Tucker, L. A. (2017). Physical activity and telomere length in U.S. men and women: An NHANES investigation. *Preventive Medicine*, 100, 145–151.

Vaiserman, A. M., & Krasnienkov, D. S., (2021). Telomere length as a marker of biological aging: State-of-the-art, open issues, and future perspectives. *Frontiers in Genetics*, 11, 630186.

Valdes, A. M., Andrew, T., Gardner, J. P., Kimura, M., Oelsner, E., Cherkas, L. F., ... & Spector, T. D., (2005). Obesity, cigarette smoking, and telomere length in women. *The Lancet*, 366(9486), 662–664.

Ward-Caviness, C. K., Kraus, W. E., Blach, C., Haynes, C. S., Dowdy, E., Miranda, M. L., & Mukherjee, B., (2016). Association of roadside air pollution with cardiovascular disease biomarkers in commuters: A panel study. *Environmental Research*, 146, 116–124.

Wen, W., Luo, S., Zhang, S., & Liang, X., (2015). The influence of circadian rhythm on the length of telomeres and the mechanism of aging. *Progress in Biophysics and Molecular Biology*, 119(2), 128–135.

Zhang, J., Rane, G., Dai, X., Shanmugam, M. K., Arfuso, F., Samy, R. P., ... & Sethi, G., (2016). Ageing and the telomere connection: An intimate relationship with inflammation. *Ageing Research Reviews*, 25, 55-69.

Zhu, Y., Xiang, Q., Xie, S., Zheng, J., & Zhang, X., (2019). Interaction between circadian rhythm and depression on telomere length and metabolic parameters in women: a cross-sectional study. *Journal of Affective Disorders*, 256, 162–168.

Chapter 8

The Organization of Urban Spaces on a Small and Large Scale Between Symmetry, Complexity, and Fractal Geometry

Gabriele Cappellato[1]
and Nicoletta Sala[2,*]
[1]Academy of Architecture, Mendrisio, Switzerland
[2]Institute for Scientific and Interdisciplinary Studies, ISSI, Locarno, Switzerland
Research Institute in Arithmetic Physics, IRFIA, Locarno, Switzerland
Institute for Complexity Studies, Rome, Italy

Abstract

This chapter examines small-scale urban organization, emphasizing garden design that incorporates architecture, aesthetics, geometry, and fractality. Three case studies cover the following topics: the regularity of form (case 1), non-orientable topology (case 2), and fractality in garden construction (case 3).

Keywords: complexity, fractal geometry, Moebius strip, symmetry, topology, urban space

* Corresponding Author's Email: nicolettasalavb@gmail.com

In: Complexity and Fractality in Human Change and Beyond
Editor: Nicoletta Sala
ISBN: 979-8-89530-575-1

1. Introduction

For centuries, the imagination of architectural forms has used the Euclidean geometric language, which translates into regular forms equipped with axes of symmetry and characterized by simple geometric constructions. In classical thought, the concept of "symmetry" was common, and Euclidean objects became an aesthetic model based on the identification of harmony and beauty with symmetry and proportion (Castellani, 2000; Sala & Cappellato, 2003; Sala & Cappellato, 2004). Therefore, it is natural that for centuries, even on an urban scale, regular geometric constructions have been used to organize space.

The purpose of this chapter is to present a brief historical description of which geometric models have been applied to the organization of urban space. It is organized as follows: The first paragraph will describe Euclidean geometry and its implications in urban planning.

The second paragraph will illustrate how fractal geometry and complexity can organize the territory and the three-dimensional representation of land.

The third paragraph will present the fractal geometry, the complexity, and their connection with the urban organization. The fourth paragraph will focus on fractal geometry for territory representation. The fifth paragraph will introduce urban organization on a small scale: garden design between architecture, aesthetics, geometry, and fractality, describing three case studies. Linear geometries, topological surfaces, and fractality will be the focus of these three projects when designing gardens and public green areas.

2. Euclidean Geometry and Urban Organization

With its regular shapes, Euclidean geometry helped the organization of urban space. The deliberate choice of a "rational" urban form led to an Euclidean geometric model imposed as an external ordering principle. The rectangular mesh is the geometric model of practical rationality that is commonly used. It attends to both practical and symbolic requirements. Regarding the practical aspect, the choice of the rectangular grid is suggested by two important factors: modularity, which allows coherent growth of the urban agglomeration, and compatibility with the agricultural land subdivision systems organized according to similar criteria.

In this way, a compromise is reached between urban growth needs and land ownership.

The rectangular mesh found wide applications in the Greek world. For example, in the Hippodamian cities such as Priene, Mileto, and Olynthos or the colonial settlements such as Selinunte and Paestum.

The term "Hippodamic" derives from Hippodamus of Miletus (4th century BC), author of a political theory and inventor of the "regular division of the city" (Figure 1), as Aristotle recalls. Two important developments of the model based on the rectangular mesh are the Roman city and the Frankish city, respectively illustrated in Figures 2 and 3.

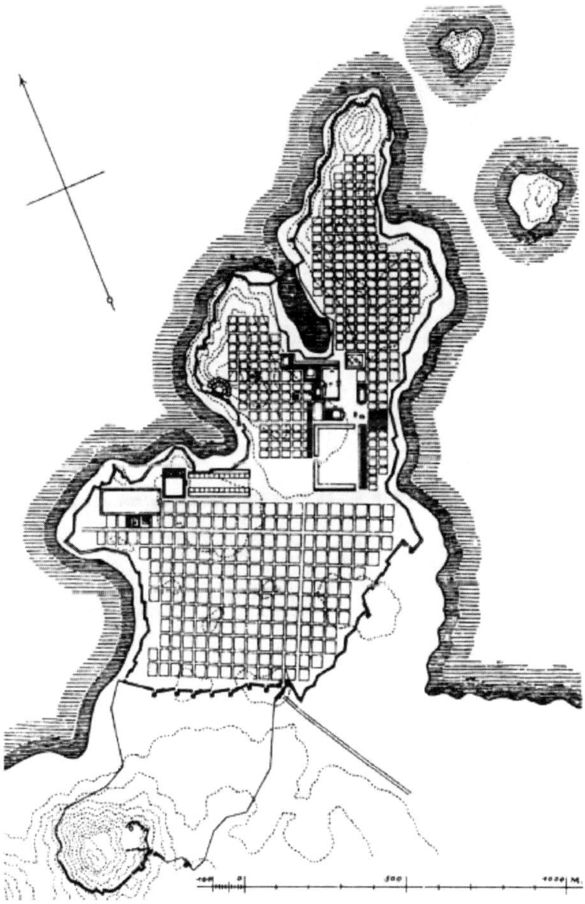

Figure 1. The regular division of the city was conceived by Hippodamus of Miletus.

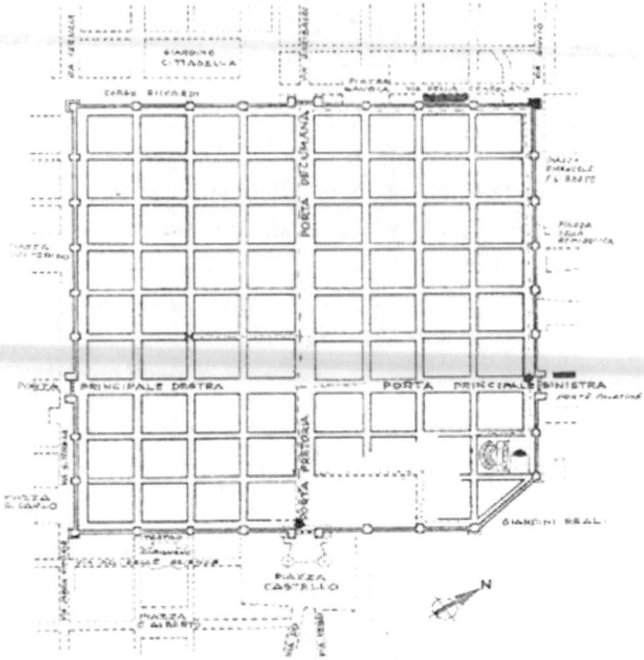

Figure 2. Example of the grid of a Roman city.

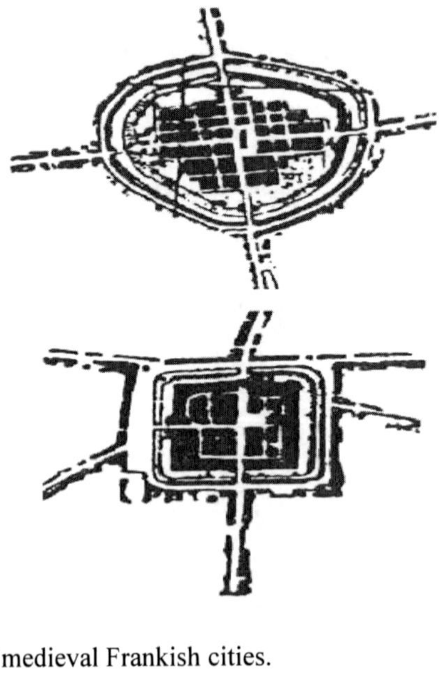

Figure 3. The grid of medieval Frankish cities.

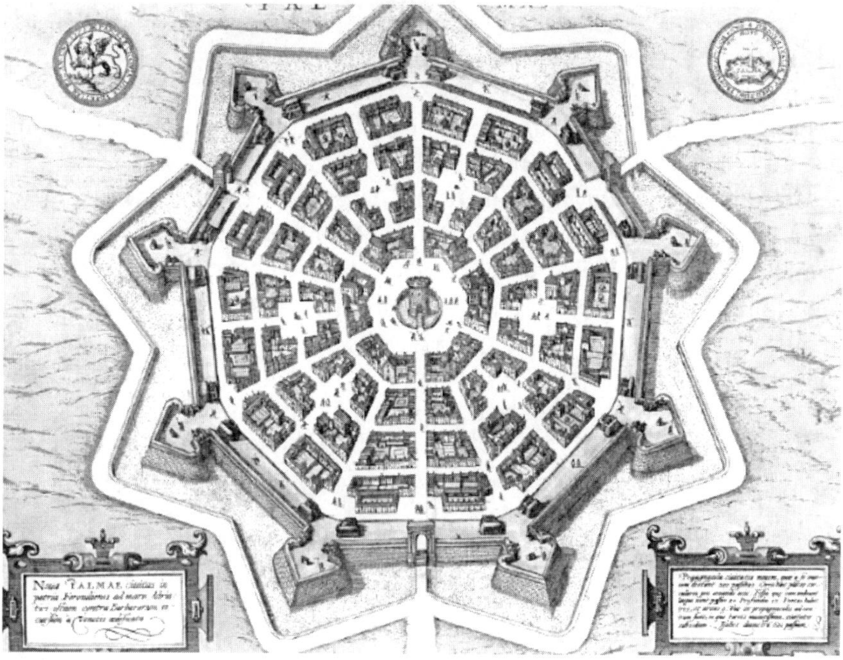

Figure 4. Palmanova is an example of a fortified city.

The Roman city was organized according to a scheme of orthogonal directions, the “cardo” and the “decumanus”; its layout is often in close relationship with the agricultural subdivision system called “centuriation.”

The medieval Frankish city was created as a result of the need for urban decentralization caused by population growth and the general improvement in living conditions throughout Europe during the 12th century.

In Renaissance urban planning, however, the polygonal model found its references in Pythagoreanism and Platonism. It is based on regular geometric shapes, such as triangles, squares, pentagons, hexagons, and octagons.

The regular polygon, with its geometric “perfection,” becomes a model of symbolic rationality. However, unlike the rectangular grid model, the polygonal model offers no margin of flexibility and adaptability to change. An example of a polygonal city is Palmanova (Friuli, Italy). It was founded in 1593 by the Venetian Senate to defend this city, the territory of the Republic of Venice, from the invasions of the Turks. Its walled enclosure has a perfect nine-sided shape, protected by the star-shaped moat; in the center, the nine-sided polygon transforms into a hexagonal square (Figure 4).

Palmanova is one of the most important models of military architecture in the modern age, with its star-shaped plan formed by three concentric city walls, two built by the Serenissima Republic of Venice composed of nine bastions and nine ravelins and a third, the most external, composed with nine lunettes, added by Napoleon at the beginning of the 19th century.

Between the fortifications, there is a dense network of tunnels, some of which are equipped and open to visitors, allowing the militias to move away from the gaze of the besiegers. Along the entire perimeter of the walls, more than four kilometers long, it is possible to practice sports or take a simple walk, admiring the majesty of the defensive works and the natural landscape rich in flora and fauna. Other notable examples of Italian fortified cities include Monteriggioni (Siena), Sabbioneta (Mantua), Civitella del Tronto (Teramo), and Corinaldo (Ancona).

3. Fractal Geometry, Complexity, and Urban Organization

The word "fractal" comes from the Latin "fractus," which means "broken, irregular." A fractal object has the characteristic of repeating its shape on different scales. This property is called "self-similarity" (Mandelbrot, 1983). Fractal geometry has interesting applications in architecture, both on a small and large scale (Sala & Cappellato, 2004).

Current urban dynamics have also allowed fractal geometry to find application in urban planning. The historical reconstruction of the processes of formation of urban structures shows that the irregular and complex shape of cities can be seen as the result of the action, prolonged over time, of multiple factors. Fractal geometry has the potential to be a tool for conceptualizing such shapes with some accuracy. Figure 5 illustrates some examples of Renaissance polygonal cities whose shape is compared with the fractal object called "Koch set" (or "Koch island"). The analogy between the shapes is interesting.

Historical analysis shows that the city tends to behave in the same way in response to similar stresses of adaptation and structural economy, growing through the progressive assembly of settlement units which present themselves, from the point of view of spatial organization and relationship, as in copies of the global organism (Donato & Lucchi Basili, 1997). We can interpret it as a type of "self-similarity" (Sala, 2006).

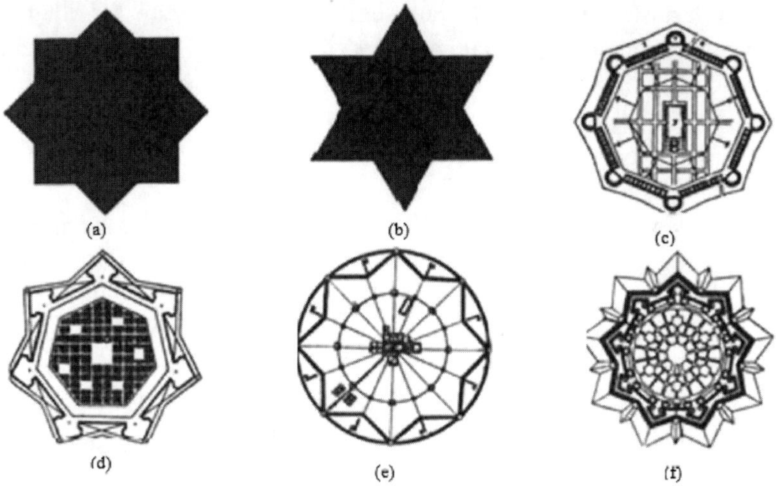

Figure 5. Fractal sets (a), (b), compared with examples of Renaissance polygonal cities (c), (d), (e), (f).

In some studies, attempts have been made to verify the potential application of fractal geometry on levels: the description of complex urban morphologies and the manipulation of design. The geometric description of real settlements in fractal terms begins with the idea that an urban aggregate is a two-dimensional geometric object in an abstract representation.

If the city is a set of built spaces, it appears as a flat area within which "voids" of various sizes can open. Suppose the density of the different voids remains constant across different dimensional scales. In that case, it can be inferred that the city can be analyzed as a self-similar object with holes with fractal dimensions between 1 and 2.

An interesting example of the application of fractal geometry in urban planning is expressed by Pierre Frankhauser, who aims to reconstruct the structure of specific fabrics rather than the global form. However, Frankhauser's model considers fractality exclusively in terms of the constancy of the average density of the built sectors at the various dimensional scales (Frankhauser, 1994; Frankhauser & Bonin, 2022). Frankhauser considers some fragments of the urban fabric of Paris and illustrates how it is possible to find fractal-type distributions even in the hierarchical structure of the road system. He saw the fractality in the French capital's urban tissue (Frankhauser, 1994).

The fractal organization of the territory is present in different cultures. For example, in the African culture (Eglash, 19999).

Ron Eglash affirms: "African settlement architecture is filled with remarkable examples of fractal structures… Traditional African settlements typically show this "self-similar" characteristic: circles of circles of circular dwellings, rectangular walls enclosing ever-smaller rectangles, and streets in which broad avenues branch down to tiny footpaths with striking geometric repetition." (Eglash, 1998, p. 21).

Some examples of urban development can be defined as complex and chaotic. They are located in the suburbs of developing countries (favelas and shanty towns). Figure 6 illustrates a Brazilian favela, a clear example of development "without any rules." Figure 7 illustrates a small Mediterranean village constructed on rock, emphasizing the complexity of its organization.

To study the evolution of a city, one can refer to the "complexity paradigm," which states that a slight variation in the initial conditions of a physical dynamic in a system in dynamic equilibrium can have large repercussions on future events.

The city can be considered as a system in dynamic equilibrium, consequently thinking of it as an organism with self-organizing capacities and highly complex autopoietic characteristics, with unpredictable behaviors that cannot be schematized through simple direct causal relationships between the variables that make up the "city system."

Figure 6. An example of chaotic Brazilian favelas.

Figure 7. A Mediterranean village shows complexity in its organization.

When simultaneously considering the city's physical, economic, and social dimensions, these realities take on a non-negligible thickness, leading to very complex processes.

A study of the city that takes into account the three dimensions mentioned cannot assume that urban growth is linear. Still, it must consider forms of local self-government that translate into relationships that regulate development through mechanisms that cannot be reproduced in classical representations of the city.

4. Fractal Geometry for the Representation of a Territory

Fractal geometry serves as a useful tool for creating 2-D and 3-D terrains and landscapes. To achieve a "fractal mountain," we can take an elastic string and apply a random vertical displacement to its midpoint. The process is repeated recursively to the middle point of every new segment (Sala et al., 2002). The random displacement decreases two times each iteration. Figure 8 shows the generation process (8a) and the mountain realized using this procedure (8b).

Realizing 3-D mountains is more complicated than realizing 2D. Some approaches are based on the midpoint displacement method, which can work with triangle, square, and hexagonal grids. Another technique that uses random midpoint displacement has been introduced by Fournier, Fussel, and

Carpenter (1982). They developed a mechanism for generating a class of fractal mountains based on recursive subdivision (Fournier et al., 1982).

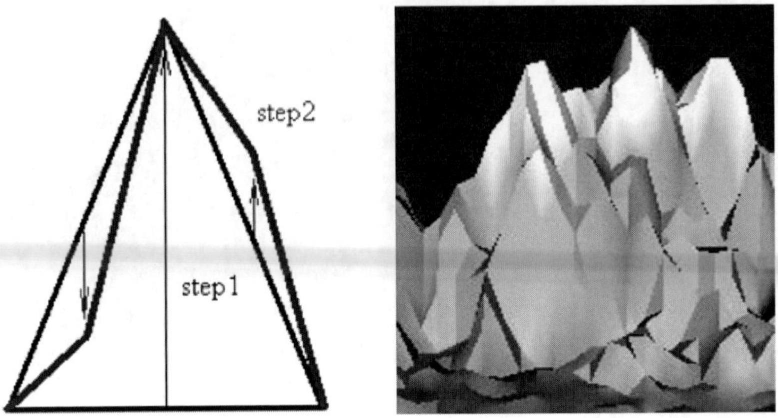

a) b)

Figure 8. Examples of fractal mountains.

For example, an initial square is subdivided into four smaller squares (see Figure 9). We obtain four points: $[x_0, y_0, f(x_0, y_0)]$, $[x_1, y_0, f(x_1, y_0)]$, $[x_0, y_1, f(x_0, y_1)]$, $[x_1, y_1, f(x_1, y_1)]$. In the first step, we add one vertex into the middle. The vertex is denoted by $[x_{1/2}, y_{1/2}, f(x_{1/2}, y_{1/2})]$.

The added vertex is shifted in a *z*-coordinate direction by a random value.

This procedure is recursively repeated for each sub-square, then on every descendant, and so on. Figure 10 reproduces a fractal terrain. The fractal dimension of this surface is $D = 2.5$.

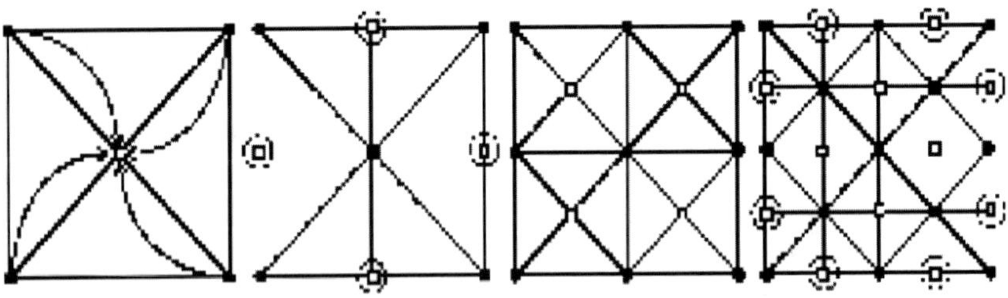

Figure 9. The first four steps of the random midpoint displacement method.

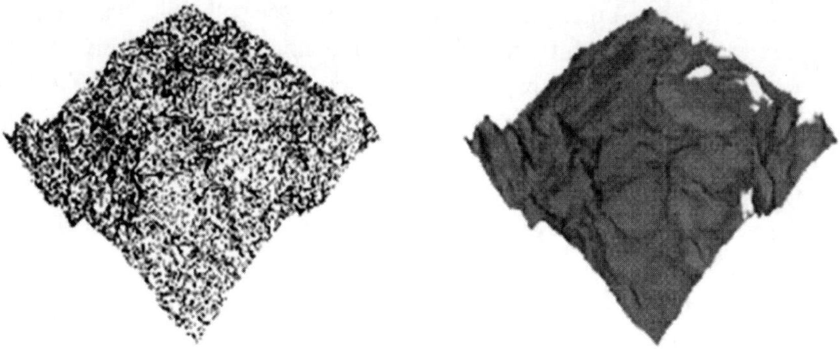

Figure 10. A fractal terrain is generated with a random midpoint displacement algorithm.

In this example, we have considered only the terrain but not its erosion (Prusinkiewicz & Hammel, 1993) and used context-sensitive rewriting processes based on a random midpoint displacement method (Peitgen & Saupe, 1988) on a triangular grid. Their method creates one non-branching river due to a context-sensitive L-system (Prusinkiewicz & Lindenmayer, 1990) operating on a set of triangles. Musgrave et al., *(1989)* introduced techniques independent of the terrain creation algorithm and can be applied to already generated data represented as regular height fields (Musgrave et al., 1989).

They introduce two methods: *hydraulic erosion* and *thermal weathering*. Hydraulic erosion is caused by water in the form of rain. Water is dropped on each vertex in a high field and transfers the material. Thermal weathering is caused by temperature changes, causing small portions of the material to crumble and pile up on the bottom of an incline (Marák et al., 1997; Musgrave et al., 1989). These models can be applied in computer graphics to generate virtual environments.

5. Urban Organization on a Small Scale: Garden Design Between Architecture, Aesthetics, Geometry, and Fractality

The taste of late 19th-century parks inspired the concept and formal construction of the garden in Europe, as these spaces represented the natural evolution of the picturesque landscapes of the previous century. Many gardens of the early twentieth century were created using green areas by replanting trees or plants from other parks, distant countries, or colonies. The garden of the eighteenth century, but also that of the following century, was divided into a series of well-marked paths, possibly tortuous, along which the visitor was encouraged to experience a series of visual and emotional sensations entirely similar to those aroused by a painting or from a theatrical scene. At specific points, the walk suddenly opened onto a suggestive panorama or revealed the unexpected presence of a pavilion or a fragment of ruin to create a particular state of mind for the visitor.

In this way, the garden was understood in its most complex articulation, giving rise to the landscape designed and built by man, a microcosm that became evidence of the way of using and transforming space and the elements of nature, varying from an English style to an Islamic paradise, to the sacred gardens of Magna Graecia, to the convent cloisters, to the formal geometric Renaissance typology, to the botanical gardens up to the places of plant experimentation or the French scenographic gardens and the Anglo-Saxon parks.

The subsequent evolutions between the end of the nineteenth century and the beginning of the twentieth century took place with the creation in particular of the "English lawn," made up of a single species of grass and used in public spaces because it was very resistant to trampling but with the need for an irrigation and constant cutting as it is subject to attacks by weeds.

The garden, in its development, structural systems, and typological characteristics, has undergone significant changes over time due to the quality and new way of life of man; this has strongly stimulated the creative research of landscape architects, especially of the twentieth century, concentrating and directing the investigation between a return to naturalism and the modern interpretation of the classic garden enhanced by the interest in exotic and native plants. In the mid-1950s, in the United States, Japan, and some European countries, particularly England and France, guidelines began to be outlined, with the help of urban planners, architects, and artists, for the redevelopment of areas intended for public gardens and parks, recovering areas devastated by war.

The new master plans envisaged the presence of extensive gardens interconnected to the existing building volumes, protection of historic parks, and protection of natural beauty, while the public green space was interpreted as a legacy of the Modern Movement and considered, in a general sense, urban or extra-urban greenery. Subsequently, in the seventies of the same century, in France, a group of architects and landscape architects developed a study plan that combined, in academic teaching, four specific disciplines:

(Anamnesis, Preparation, In-depth reading, Relative thinking) which outlined a new and unprecedented concept for the design of gardens and parks.

Anamnesis or Reminiscence consisted of analyzing the signs of time that had stratified on the site and shaped a specific landscape, identifying the vocations of the territory they wanted to protect and transmit.

The preparation considers the landscape and the interpretation of the site as an ongoing process, enhancing the natural elements: the cycle of the seasons, the climates, the processes of alternating day and night, and the changes in time and life.

In-depth reading or in-depth vision was a critical alternative to the plan's vision, inducing the recognition of the qualitative perception of space and the environment, extending the in-depth intentionality of the project to the components of the landscape and the site, and integrating form and function.

Relative thinking consisted of examining the site and the landscape beyond its borders and examining the relationship between buildings and peripheral areas (roadsides, uncultivated areas, non-places).

These conceptions have directed and urged contemporary garden designers to adapt and follow designs and compositional aggregations often structured by disharmonious geometries concentrated on a single central axis or articulated in diagonal and asymmetric multi-axial spaces.

Today, the garden project is seen as a space meant for socializing and bringing people together, especially in light of the significant changes that have taken place in human society and the environment.

Using new materials due to technological development, such as plastics, metals, rock formations, glass wool, fiber optic lighting, and artificial fog, represents the most widespread components of a new contemporary garden or public park project.

The latter reveals itself as an ecological regenerator of interstitial spaces, abandoned quarries, landfills, motorway rest areas, or as a non-place. The designer who intervenes in these areas is understood as the "builder" of a new landscape, a bearer of modernity who does not forget to understand and share that the past is an essential potential for the future since it traces its history and traditions. In this new project vision, the landscape designer brings out nature's potential in promoting the natural ventilation needed by the building built around it, establishing a connection and a link of continuity within the area.

The constructive element of interconnection between outside and inside can be the open portico or a terrace, a platform connected to the building through a mobile or light canopy. The water collection for its supply and disposal must be done through tanks and gargoyles, while light can be used as a compositional factor of the architecture to make the space fluid and bright. Built with white walls that favor light diffusion, it creates its structure in a continuous dialogue between architecture and territory. Today's landscape can be seen as the semantic equivalent of natural beauty, a system of physical, biological, and anthropic aspects that characterize a specific geographical site shaped over time by the conscious and systematic action of the community that inhabits it.

The following three case studies show new forms of protection and culture. They have different sizes, metric scales, qualities, expressive languages, and approach strategies. Within this thought, they underline the topic's relevance by presenting an interesting choice of solutions linked to mathematics, fractal geometry, and the signs that structure and compose them.

5.1. Case 1. The Regularity of the Form: Caille Garden, Lyon, France (Michel Desvigne - Christine Dalnoky, 1992)

The United States neighborhood of the city of Lyon, France, designed by the architect Tony Garnier, is the initial idea of reference for this garden, which draws inspiration from the urban configurations of the road network and the compositional configurations of the aggregation of the blocks created in the first thirties of the twentieth century. A geometric series of parallel bands characterize the garden of plants, small avenues, benches suitable for games, and clearings, creating regular but always visible private corners.

The space created unifies the paved nature, in local red stone, for the extended surface of the square, in contrast with the vegetal, more intimate and relaxing aspect of the garden combined with a planting of pines, rhododendrons and laurels that relate to the geometry modular and rigorous of the built elements such as the concrete benches, the granite paths or the main paths in cast-in-place concrete.

In continuity with the geometric layouts of the neighborhood, the project plan comprises various parallel bands to protect the central space with various linear concrete elements that delimit the different places. The latter is characterized by an essential aesthetic, where the two large hedges of shrubs, pruned as regular volumes, run along the longer sides of the area concerned, flanked by two other smaller hedges which are found in the center of the garden together with the row of pre-existing plane trees that are linked to the planting of pines.

Adopting a mixed garden and square intervention allows a sort of mediation between the desire for a modern expression of the proposed all-green space and the rigid Hausmannian grid of public spaces dedicated to squares, streets, and gardens.

5.2. Case 2. Non-Orientable Topology: Möbius Garden, Adriaan Geuze West 8 (Het Gooi, Holland, 1998)

The garden called Möbius is the continuation of the virtuous play of connecting the lines present in the architecture of the house of the same name, designed by UN Studio in 1998 and stands on a part of the adjacent forest.

The design of the house presents, in the ring of spatial layout, pointed angular lines conceived as an intertwined and continuous trajectory, which aligns all the functional spaces of the life of a highly personalized family.

The idea refers to the Möbius strip or the Möbius loop. This one-sided surface can be constructed by joining the ends of a rectangular strip together after having rotated one of the ends by half (Figure 11). This surface has interesting properties, such as having only one side and remaining in one piece when split down the middle. As a mathematical object, it was discovered by Johann Benedict Listing (1808-1882) and August Ferdinand Möbius (1790-1868) in 1858. It consists of a plan with axes that intersect to form a double spiral generated by a ring of two interconnected lines that integrate the three fundamental components of the project: program, circulation, and structure.

These main axes that structure the non-orientable surface band represent the materials used in architecture and the garden: glass, concrete, stone, water, and plants.

Figure 11. Möbius strip: an example.

The intense green surface of the garden is conceived as a means of connection between the house and the vegetal element of the forest, and its design takes up the same concept of the seamless topological ribbon.

The presence of articulated and intertwined passages and paths was deliberately introduced to interrupt some architectural elements of the house with a very well-defined and refined appearance, such as the greenhouse, the functional connections, the service spaces, and the swimming pool.

The latter is black, and its horizontal body of water reflects the context of the trees and the sky, making them iridescent colors around it and accentuating the reflection on the glass façade of the house. The winding grass paths connect the various levels of the garden; they turn, intertwine, and go back before disappearing into the dense undergrowth, while the textures of the wood and the seriality of the glass surfaces with the changing of the seasons are fixed in this work of architecture and landscape called Möbius.

5.3. Case 3. Fractality in Garden Construction. Botanical Garden and Park in Montjuich, Barcelona, Spain, 1999 (Carlos Ferrater, Josep Lluis Canosa, Architects; Bet Figueras Landscape Designer; Artur Bossy, Horticulturist; Joan Pedrola, Biologist)

The project involves the creation of a new botanical garden for the city of Barcelona, specifically focused on the flora of the climatic regions of the Mediterranean, and was developed through dialogue, comparison, and intertwined work by various specialists involved (architects, landscapers, horticulturists, and biologists). The concept of the work starts from the desire to create a project that could allow the place to self-qualify by generating a new form of territory from its own morphological and topographic conditions.

This unique condition inspired the concept of developing a geometric system resulting from the arrangement of a triangular grid. This grid adapts to different tree species by organizing them according to the distinct ecosystems, forming a shape that reflects the various directions of paths and existing buildings.

This grid can be superimposed and adapted to the terrain when it increases or decreases in surface area based on the orographic slope.

The grid positioning aligns with the three principal directions of the topographic lines (height, width, depth), ensuring that at least two vertices of each triangle are always at the same level and without slope.

This construction system consists of a series of double concave or convex walls that can be superimposed and adapted to the terrain, organizing and designing the landscape in an order and a fractal dimension starting from the irregularity and fragmentation of the surfaces of the territory.

In fact, by slightly moving the ends of the triangles, the grid breaks down, and the landscape becomes multifaceted in every part of the surface, reaching a precise singularity of orientation and slope.

The other parts of the grid maintain the same structural shape because they respond to the needs of accessibility, functionality, and planting and are regulated by a variable geometric order depending on the position of the site and the vital needs of the garden, which are the service infrastructures, the drainage and water supply.

The planting of each single plant element balances the excessive virtuality, bringing a concrete formal order to the construction of the garden and allowing the fractal geometric dimension of the form to be incorporated with the vegetal presence of nature in a synthesis between ecological balance and artificiality of the built infrastructures.

In this way, the project's geometric device achieves its most important objective: ensuring regular control over the garden's morphology.

Conclusion

The term "territory" encompasses various interpretations and descriptions that architects, urban planners, geographers, and mathematicians use to articulate their ideas. It serves as a primary resource and the center of complex material processes, existing as a physical sphere and a limited asset. Organized as a syntagm, it features several fundamental elements, including altitude, atmosphere, water, sea, rocks, vegetation, fauna, and human activities.

It constitutes the consequence of formation phenomena and continuous production of results from anthropic, natural, historical, cultural, economic, and social structural elements contributing to its formation and transformation. This complex composition requires an assumption of responsibility and moral quality to respond to the needs of its components expressed in the formation of an adequate awareness of the place.

The territory is an asset to be protected, especially in the relationships between phenomena united by functional links but also by symbolic values that each of us possesses because it represents the image of a human landscape imbued with connections between environments, cultures, opportunities, and new potentials that characterize human life and which are always organized around pre-existing environmental elements.

The crisis of identities, the change in social relationships, and the negative environmental transformations show today a scenario whose alternatives and aspects that must be highlighted force us to act with a view to large and far-sighted projects so as not to fall into an individualistic micro-nationalism.

The references and examples that we have reported and selected are different from each other in terms of the type of approach, sizing, metric scale, quality, expressive language, and strategies, but within this thought, they underline the relevance of the topic addressed which presents itself in an interesting choice of solutions linked to mathematics, fractal geometry, the signs that structure them and the areas that compose them. What emerges for everyone is the search for a solution that is not only superficial and formal concerning the environmental problem but also a different desire for a method of transforming the space investigated.

We believe that today, architecture is understood as a discipline at the service of humanity, shaped by the nature and typology of the place and projected towards a healthy and better civil society that is respectful of the environment and attentive to the diffusion of solidarity networks.

As Donato and Lucchi Basili (1996, p. 17) state: "Fractal geometry seems to open the way to the possibility of overcoming the rational/organic dichotomy, to the extent that it can offer the key to a rigorous description of the structural properties of unplanned settlement forms, with a view to their assimilation and design re-elaboration."

Fractal geometry today serves as a vital tool for designers to express attentiveness and sensitivity, ensuring that every change occurs without distorting the deep authenticity of a place. Only in this manner can the territory evoke strong suggestions and sensations that encourage participation by presenting an image of a landscape that transcends immediate emotion.

References

Castellani, E. (2000), *Simmetria e natura*, Bari: Laterza.

Donato, F., & Lucchi Basili, L. (1996). *L'ordine nascosto dell'organizzazione urbana.* Milano: Franco Angeli.

Eglash, R. (1998). Fractals in African settlement architecture. *Complexity*, *4*(2), 21–29.

Eglash, R. (1999). *African Fractals*, London: Rutgers University Press.

Fournier, A., Fussel, D., & Carpenter, L. (1982). Computer Rendering of Stochastic Models. *Communications of the ACM*, 25, 371–384.

Frankhauser, P. (1994). *La Fractalité des Structures Urbaines*, Paris: Anthropos.

Frankhauser, P., & Bonin, O. (2022). Fractalopolis–Sustainable planning for metropolitan areas by a scaling approach. *Nobel X conference in Urbanism 2022, Urban futures in the face of overwhelming uncertainty.*

Mandelbrot, B. (1983). *The Fractal Geometry of Nature*, New York: W. H. Freeman.

Marák, I., Benes, B., & Slavík, P. (1997). Terrain Erosion Model Based on Rewriting of Matrices. *Proceedings of WSCG-97*, II:341-351.

Musgrave, F., Kolb, C., & Mace, R. (1989). The Synthesis and Rendering of Eroded Fractal Terrains. *Computer Graphics*, 23(3):11-1-11-9.

Peitgen, H., & Saupe, D. (1988). *The Science of Fractal Images*. New York: Springer-Verlag.

Prusinkiewicz, P., & Lindenmayer, A. (1990). *The Algorithmic Beauty of Plants.* New York: Springer-Verlag.

Prusinkiewicz, P., & Hammel, M. (1993). A Fractal Model of Mountains with Rives. *Proceedings of Graphics Interface '93*, 30(4):174-180.

Sala, N., & Cappellato, G. (2003), *Viaggio Matematico Nell'Arte e nell'Architettura*, Milano: Franco Angeli.

Sala, N., & Cappellato G. (2004), *Architetture della Complessità. La geometria frattale tra arte, architettura e territorio*, Milano: Franco Angeli.

Sala, N. (2006). Fractal geometry and architecture: some interesting connections. *WIT Trans. Built. Environment, 86*, 163–173.

Sala, N., & Cappellato, G. (eds.) (2018). *Chaos and Complexity In the Arts and in Architecture: Research in Progress*. New York: Nova Science Publishers.

Sala, N., & Cappellato, G. (2019). *Connections between Mathematics, the Arts, and Architecture,* New York: Nova Science Publishers.

Sala, N., Metzeltin, S., & Sala, M. (2002). Application Of Mathematics in the Real World: Territory And Landscape. *Proceedings International Conference Mathematics for Living,* 333 – 336.

Sala, N. (2002). The presence of the Self- Similarity in Architecture: Some examples. In M. M. Novak (ed.), *Emergent Nature*, Singapore: World Scientific, 273 – 283.

Chapter 9

Is It Possible to Optimize Psychophysical Complexity in Élite Athletes? Photoplethysmography in Sport Psychology

Francesca Stecchi[1,*]
and Laura Cancellara[2]
[1]Founder Insynergy MindLab, Padua, Italy
[2]PM Mind Force Society/MIUR, Rome, Italy

Abstract

This study aims to develop an intervention protocol to assist professionals in sports psychology in managing pre-competition mental preparation and post-training recovery following intense sessions. The target group for this study consists of professional and elite athletes. The work focused on using a non-invasive monitoring technique, photoplethysmography (PPG), which measures variations in blood volume in tissues using optical sensors placed on the right and left index fingertips. Alternatively, these sensors can also be positioned on the toes. These sensors emit infrared LED light that penetrates the tissue, detecting changes in reflected or transmitted light caused by blood volume fluctuations associated with heartbeats. Each heartbeat momentarily increases blood volume in the vessels, altering light absorption and generating a signal that the PPG registers and analyzes. Through this device, various functional parameters—detailed in subsequent sections—were detected, and PPG was paired with biofeedback. Through training and formative experience in sports psychology, we integrated data obtained from the PPG device, considering the complexity of

* Corresponding Author's Email: francesca.stecchi@insynergy.it

In: Complexity and Fractality in Human Change and Beyond
Editor: Nicoletta Sala
ISBN: 979-8-89530-575-1

blending advanced physiological monitoring with the psychophysical management of athletes. Some data were then discussed, highlighting the importance of a shared, multidisciplinary approach with other team professionals, such as coaches, sports physicians, osteopaths, and physiotherapists, to work synergistically. The study focused on heart rate variability (HRV), vagal activity, autonomic nervous system (ANS) monitoring, and biofeedback, adopting a holistic approach to managing elite athletes. Thirty-two professional athletes aged 16 to 30 from sports including tennis, volleyball, and sprint canoeing were evaluated to assess the autonomic nervous system (ANS) function, recovery capacity, and emotional states at rest and under stress following intensive training. Additionally, the study explored the relationship between Medically Unexplained Symptoms (MUS) and physiological markers. Results showed significant correlations between vagal activity, recovery times, and the emotional states of male and female athletes. Athletes with higher vagal activity demonstrated a more effective ability to adapt to stressors, greater energy, and enhanced recovery. Moreover, MUS was linked to specific changes in HRV patterns and vagal activity, suggesting that integrating non-invasive monitoring with a multidisciplinary approach could provide new tools to optimize both performance and overall well-being in professional athletes.

Keywords: psychophysical complexity, elite athletes, photoplethysmography, sports psychology, body awareness, emotional awareness, antifragility

1. Introduction

In the international context of elite sports, focusing on enhancing mental abilities and improving athletes' psychological well-being has become essential for optimizing high-level sports performance. Emerging scientific perspectives emphasize the need to prioritize psychological health alongside physical health.

The sports psychology intervention programs currently in use also aim to provide athletes with optimal resources for improving emotional control under pressure. Elite athletes are tested in various circumstances, and the factors affecting their success frequently extend beyond sports. Over time, research has adopted a holistic perspective on the athlete, whose subjective experience is divided into multiple, interwoven levels. An athlete is first and foremost, a person, before being an athlete (Stecchi et al., 2022).

In this vision, a deep understanding of mind-body dynamics, implementing sport psychology interventions, and focusing on managing stress and emotions are fundamental pillars in supporting athletes to achieve peak performance. Focusing only on physical performance aspects, biometric and physiological parameters, tactical-specific training techniques, and outcome objectives involves a reductionist and one-dimensional approach, concentrating almost exclusively on the physical and performance aspects. Performance is no longer just a final result indicator achieved by the athlete; it expresses a broader, holistic, and systemic vision that considers complex, interacting multidimensional factors. Psychological resilience, antifragility processes, overall psychophysical health, contributions from technical and specialized staff, intervention context, life cycle phases, and other dimensions intersect on various levels, dynamically influencing each other.

Therefore, high-level performance is the product of complex physiological, psychological, environmental, and social interactions. The optimal management of elite athletes requires an approach considering these interactions' dynamic and non-linear nature (Davids et al., 2014).

Although effective in enhancing specific physical, technical, and tactical abilities, traditional training approaches often overlook subtler but equally important aspects, such as psychological factors like stress, the effects of insufficient recovery, dysfunctional routines, and the impact of emotional states and psychological well-being on performance. This study addresses the need to develop a fast and precise protocol that can be applied not only during the training season but also, and especially, before significant competitions like the Olympics. This approach allows for targeted interventions, focusing support activities on the most functional aspects for each athlete in real-time, optimizing resources, and reducing the adaptation time necessary to maximize performance.

The study further aims to explore an integrated model combining advanced physiological monitoring with a multidisciplinary approach to ensure elite athletes' physical and psychological well-being. Recent research in sport psychology highlights the growing importance of advanced technologies like biofeedback, neurofeedback, and PPG to precisely monitor the autonomic nervous system (ANS) responses in elite athletes. These tools enable the detection of the balance between the sympathetic and parasympathetic systems, essential for understanding stress and recovery states in athletes (Aubert et al., 2003). Studies such as that by Bellenger et al., (2016) demonstrate that high HRV is associated with improved adaptive capacity and greater resilience to stressors, allowing athletic performance optimization through targeted, personalized interventions.

The autonomic nervous system (ANS) plays a crucial role in regulating stress response, emotional tone, and physiological recovery. The balance between the ANS's sympathetic and parasympathetic branches is essential for effectively modulating the response to intense loads and promoting optimal recovery (Aubert et al., 2003). In this framework, heart rate variability (HRV) emerges as a sensitive indicator of ANS status, capable of detecting even minimal alterations in the organism's adaptive capacity. Elevated HRV is often correlated with better autonomic balance and greater stress resilience (Bellenger et al., 2016), indicating an enhanced ability to maintain physiological stability and recovery.

This study addresses the need to operate in elite contexts by utilizing a rapid measurement (with an effective duration of 5 minutes) that provides accurate data on athletes' psychophysical state and sleep-wake rhythms, allowing for timely, targeted, and precise interventions to optimize well-being and, consequently, performance. Additionally, this protocol has proven effective in promoting a multidisciplinary approach, integrating the work of coaches and the entire staff—including sports physicians, physiotherapists, osteopaths, nutritionists, and physical trainers—in a synergy aimed at the athletes' well-being and peak performance.

In the following sections, we will see how PPG has benefited our work as sports psychologists in creating a monitoring protocol useful for pre-competition work with elite athletes.

By monitoring autonomic nervous system (ANS) parameters, photoplethysmography (PPG) reveals crucial information about sleep disorders through heart rate variability (HRV) analysis, particularly via the low-frequency (LF) and high-frequency (HF) components. Recent studies (Park et al., 2022) demonstrate that HRV measured through PPG can provide data comparable to ECG in detecting autonomic tone alterations and sleep disorders. This is particularly important for evaluating respiratory alterations and measuring vagal tone variations, both essential for nighttime recovery and stress management in athletes.

Specifically, the high-frequency (HF) band of HRV is closely associated with parasympathetic vagal activity, which increases during deep sleep and serves as a key indicator of nighttime recovery quality. Since it is higher during deep sleep and rest, an increase in HF during sleep indicates good recovery and strong vagal activation, suggesting a greater ability of the body to relax and regenerate. Conversely, if HF parameters are high during the day, it indicates a reversed circadian rhythm (night and day) and that the athlete did not recover overnight. An athlete not aligned with the sleep-wake cycle will be less focused, more prone to errors, obviously fatigued, and at greater risk of injury. On the other hand, if HF values are low at night, this could indicate poor recovery or difficulty reaching a deep sleep phase, often associated with stress or physical overload. These parameters, therefore, allow for the assessment of not only the duration of sleep but also its quality and, consequently, the effectiveness of recovery for the athlete.

2. Theoretical Background

The biopsychosocial approach and salutogenic perspective are gaining increasing relevance in managing the psychophysical performance of elite athletes, opening new horizons for sports psychology. This paradigm shift, supported by recent discussions from the International Olympic Committee (IOC) (Reardon et al., 2019), represents a renewed view of the athlete: no longer viewed solely as a "performing body" but as an entire individual whose psychophysical health depends on a dynamic balance among physical, mental, and social dimensions (Reardon et al., 2019). This holistic perspective emphasizes mind-body unity, essential for embodied performance. Adopting an integrated approach to psychophysical well-being requires the involvement of highly specialized professionals, including sports psychologists, whose specific multidisciplinary training represents a key element. This professional role, equipped with specific clinical and technical skills, supports athletes in managing competitive pressures, and optimizing mental, and physical capacities through evidence-based, scientifically validated interventions. The psychologist's role extends well beyond essential psychological support to facilitate cooperation with coaches, nutritionists, and other team members, promoting a synergistic support network that fosters the athlete's overall well-being (Gustafsson et al., 2015; Giges & Petitpas, 2020).

Within this framework, intervention strategies following the principles of matching and tailoring are advisable, allowing for therapeutic methods to be adapted and personalized to each athlete's psychological characteristics and unique needs. This personalization process involves a thorough assessment of the athlete's individual history, competitive environment, and the specific demands of their sport. By adopting these tailored approaches, it is possible to ensure not only the effectiveness of strategies but also their sustainability, maximizing athlete well-being at each stage of their competitive journey (Fletcher & Sarkar, 2012).

An in-depth understanding of the biopsychosocial paradigm and the application of a salutogenic approach represents a fundamental step forward for sports psychology. These practices promote balance between the psychophysical performance components and shift the focus to burnout prevention and resilience building, enabling athletes to face the challenges of their careers with greater awareness and strength.

In summary, the paradigm shift highlighted by the IOC discussions and current scientific research underlines the need to consider the athlete's biopsychosocial complexity. The specific training of sports psychologists and the implementation of personalized interventions, through matching and tailoring, offer a unique opportunity to integrate advanced psychological practices into sports preparation, fostering a culture where mental health becomes integral to physical health and optimal performance.

In light of the aforementioned background, our theoretical-methodological proposal emerges. We believe that implementing "integrated protocols" in sports psychology — focusing on the psychobiological assessment of elite athletes, psychophysical performance monitoring, and the correlation with emerging complexity factors—is the intervention path closest to the real needs of diverse work contexts.

Implementing personalized intervention protocols that consider elite athletes' global psychophysical performance can be essential to maximizing athletic potential and optimizing outcomes. In professional sports, even the slightest improvement can distinguish between excellent, optimal, or ordinary performance levels. Elite athletes face complex physical and psychological stress levels, and their success also depends on the optimal balance of physical strength, mental resilience, psychological resilience, and recovery capacity. The article subsequently proposes a reflection aimed at analyzing, exploring, and understanding the "resilient complexity factors" that optimize both physical capacities related to sports performance and psychological capacities—two inseparable dimensions in high-level sports performance that work synergistically to build an optimal state of psychophysical and sports well-being. To maximize performance, developing a plan that integrates the athlete's specific needs, the characteristics required by the specific sport, and the needs of different training phases is essential. We believe it is helpful to focus more attention on aspects that significantly influence sports performance and, more generally, the athlete's quality of life.

This work aims to focus on:

- Integration between Wellness Sciences and Sports Psychology: how experimental and innovative techniques meet in the field;
- Supporting the athlete's psychophysical performance;
- Increasing awareness: body-kinesthetic/self-awareness;
- Managing state stress;
- Monitoring the sleep-wake rhythm;
- Understanding somato-emotional reactivity;
- Promoting recovery times and energy restoration.

3. Methodology

3.1. Participants

The study involved 32 professional athletes from three different sports: tennis (n = 3), volleyball (n = 26), and sprint canoeing (n = 3), with an equal number of male and female participants (16 each). Selection was based on participants' availability to follow the monitoring protocol over the entire two-month study period. All participants were either currently or previously supported by us in sports psychology within various Federations or private practices.

3.2. Instrumentation

Physiological monitoring was conducted using a non-invasive medical device employing multichannel photoplethysmography technology to measure parameters related to HRV and autonomic nervous system (ANS) activity. This device provided precise, continuous data on HRV, vagal activity, and parameters related to recovery, stress management, and emotional regulation. Specifically, we examined parameters related to average heart rate, sympathetic and parasympathetic activity, the interaction between the two systems, and heart rate variability.

3.3. Procedure

Each athlete underwent a preliminary evaluation, including a structured interview and questionnaires to identify the presence of medically unexplained symptoms (MUS). Physiological monitoring was conducted at three different stages:

- T.0. Resting: Measurements were taken at complete rest to establish baseline ANS activity parameters.
- T.1. Post-training: Measurements were immediately taken following an intense training session to evaluate acute psychophysical stress response.
- T.2. Recovery: A further measurement was taken 30 minutes post-training to assess autonomic recovery dynamics.

Each measurement session lasted 5 minutes, during which continuous data on HRV and sympathetic-parasympathetic balance were collected.

3.4. Analyzed Parameters

The primary physiological parameters analyzed included:

- Heart rate variability (HRV)
- Vagal activity
- Sympathetic-parasympathetic balance
- Recovery index: Based on the speed of returning to baseline ANS values after exercise
- Stress index: Representing the immediate autonomic response to intense training
- Correlation with MUS: (a questionnaire was administered to assess vague and non-specific symptoms). The analysis was conducted on the correlations between MUS reported by athletes and variations in physiological markers, using a comprehensive view of data, results, and clinical interviews with athletes.

4. Results

4.1. HRV, ANS, and Vagal Activity Analysis

The analyses revealed significant differences in HRV, ANS activity, and vagal patterns across the three measurement phases. Athletes with higher vagal activity demonstrated a greater capacity for rapid recovery from intensive workloads, suggesting more effective autonomic modulation. Additionally, Bellenger et al., (2016) observed that athletes with higher resting vagal activity improved recovery time and stress management induced by training, indicating more effective autonomic modulation. This type of recovery proved crucial for endurance performance and reducing overreaching or overtraining effects.

4.2. Performance Correlations

A significant correlation was found between the post-training recovery index and athletic performance in subsequent weeks. Athletes with higher vagal activity and more efficient autonomic modulation showed greater consistency in high-level performance, highlighting how effective recovery management is key to maintaining long-term performance quality. One of the athletes in the study won a silver medal at the Paris 2024 Olympics.

4.3. MUS and Physiological Indicators

Athletes reporting high levels of fatigue, sleep disturbances, or frequent muscle pain showed a reduction in vagal activity during recovery phases. These findings suggest that medically unexplained symptoms (MUS) may impair the autonomic nervous system's ability to recover effectively after intense training. Additionally, the research highlights how low sleep quality and muscle pain can alter physiological markers, such as HRV, reducing the adaptive capacity of the autonomic nervous system and negatively affecting the athlete's recovery post-training. This aligns with the findings of Nevill et al., (2021), which examined recovery and sleep practices in elite athletes.

Their study demonstrated that sleep disturbances and fatigue could hinder the autonomic nervous system's ability to restore effectively after intense effort, with reduced vagal activity associated with higher stress levels and recovery difficulties.

4.4. Adaptation and Resilience Indicators

Longitudinal analysis revealed that athletes showing greater variability in physiological parameters across training sessions tended to achieve significant long-term performance improvements. This phenomenon aligns with the concept of resilience and, more importantly, antifragility, suggesting that exposure to variable stressors can promote positive adaptations. Unlike resilience, which implies the ability to return to the initial state after an adverse event, antifragility goes further: It represents a system's ability to improve, evolve, and grow through exposure to unpredictable and variable events, extracting the best from these events and adapting accordingly.

5. Discussion: The Complexity of Non-Invasive Monitoring in the Physical and Emotional State of the Athlete

Non-invasive monitoring of physiological parameters, such as HRV and vagal activity, represents an essential tool for real-time understanding of athletes' physical and emotional states. HRV variations can reflect physical, mental, and emotional stress levels, providing valuable information for adjusting training loads or implementing psychological support measures. Advanced technologies allow for early identification of autonomic imbalances before they manifest as injuries, performance reductions, performance anxiety, or excessive stressors. Aubert, Seps, and Beckers (2003) observed that heart rate variability (HRV) is a key indicator of physiological stress response and the balance of the autonomic nervous system in athletes.

HRV is heavily influenced by training but also by emotional states and can vary significantly based on the athlete's physical conditioning, recovery status, and stress levels. Regular HRV monitoring through photoplethysmography devices offers valuable insights into the state of the autonomic nervous system (ANS), allowing early detection of fatigue or overload. This monitoring can help sports psychologists select the most appropriate recovery and stress management techniques to optimize mental and physical preparation, enabling athletes to be more aware of their body's signals and emotions and, thereby, better able to manage them and recognize when to pause, reducing injury risk or sports burnout.

Continuous, non-invasive physiological monitoring fits well into a multidisciplinary approach, where sports physicians, nutritionists, physiotherapists, and psychologists work together to optimize athlete management. Monitoring data can be used to customize recovery strategies, improve diet, and prevent injuries or training overload. This specialist collaboration enables the analysis and care of the athlete as a complex system, integrating different areas of intervention.

A key aspect of this holistic approach is remembering that, before being athletes, individuals have unique physical, psychological, and emotional needs. Recognizing the importance of supporting the athlete's overall well-being, beyond performance optimization, is essential for ensuring a long athletic career and minimizing or eliminating injury and burnout risks. Continuous monitoring of physiological state enables athletes to become more aware of their health, emotions, and functioning. Clarsen, Myklebust, and Bahr's (2013) work has been fundamental in developing methods for identifying and measuring overuse injuries in athletes. Their contribution highlights the importance of specific assessment tools that allow for the identification of both physical fatigue symptoms and physiological stress. In our work, using non-invasive tools like photoplethysmography provides real-time data on the athlete's psychophysical state, allowing for personalized preventive interventions for individuals and teams.

This work with athletes emphasized body and emotional awareness based on device data. Through a dynamic and targeted approach, it was important to also work on antifragility. Some athletes not only adapt but thrive under pressure, strengthening their recovery capacity and growth in response to stress and competition challenges. Educating athletes and staff on the importance of mental preparation and psychological well-being is fundamental to achieving high performance.

6. Psychological Performance Programs: Benefits and Complexity

The implementation of advanced strategies in sports psychology allows for personalized support tailored explicitly to elite athletes. Techniques such as biofeedback, mindfulness, and digital tools for physiological parameter measurement improve self-awareness and enable timely interventions in situations that cause distress and fatigue. Biofeedback technology (BFB) presents significant opportunities for sports psychology, especially in training psychological skills. The goal is to reduce stress in competitive situations and enhance athlete performance (Blumenstein et al., 2017). Recent studies have shown positive results of BFB training in various sports, demonstrating its effectiveness in enhancing emotional regulation and psychological resilience (Blumenstein et al., 2017).

Considering complexity factors and the inherently dynamic and non-linear nature of athletes' life events (Bergeron et al., 2024) also means monitoring and evaluating the perturbing aspects of the psychophysical performance process. For example, for young elite athletes participating in the Olympic Games (the IOC mentions adolescents with asynchronous development of physical, physiological, psychological, and social characteristics continuing until early adulthood), factors increasing global complexity in highly competitive environments may emerge as vulnerability and criticality factors. This refers to the need to explore "emerging" complexity factors that, if studied, contribute to injury prevention and can also be expressions of maladaptive psychological factors that negatively influence physical preparation and on-field performance (such as stress and mental fatigue).

6.1. Managing State-Stress

State stress, or situational stress, influences an athlete's emotional reactivity during competitions. A study published in *Frontiers in Psychology* by Tossici, Zurloni, and Nitri (2024) analyzes the dynamics of allostasis (physiological equilibrium in response to external factors) and eustress (positive stress) versus distress (negative stress) in sports, highlighting the importance of multidisciplinary protocols to monitor and manage athlete stress, with direct benefits on performance and mental well-being. Figure 1 highlights the intersection of predictive and risk factors involved in managing *State-Stress* in elite athletes.

The factors include:

a) *Performance Optimization*: High levels of *State-Stress* can lead to anxiety and compromise cognitive functions essential for athletes during competitions. Effective stress management can enhance concentration, reaction time, and decision-making abilities under pressure.

b) *Psychological Health*: Chronic exposure to high stress levels may negatively impact an athlete's mental health, generating dysfunctional emotional experiences (e.g., anxiety/depression). Elite athletes often face intense emotional tension during competitions, making them particularly vulnerable and/or fragile.
c) *Physical Well-Being*: State stress can also have physiological effects, such as increased heart rate and muscle tension, impairing athletic performance. Stress management techniques can help monitor and maintain optimal physical conditions over time.
d) *Long-Term Success*: Learning to manage *State-Stress* effectively can contribute to an athlete's career sustainability, helping them avoid burnout and related injuries, such as those stemming from overwork and mental fatigue.

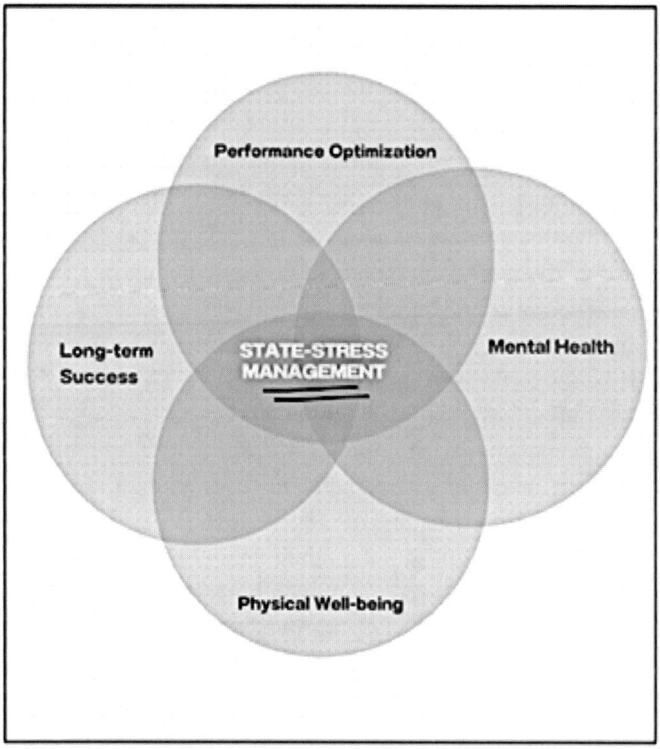

Figure 1. Management of State-Stress.

In summary, learning to manage stress is essential to prevent intense emotional experiences, which generate anxiety or nervousness, from interfering with performance. Strategies such as progressive relaxation, controlled breathing, and positive self-talk help athletes maintain calm and perform their best skills effectively.

7. Emotional Dysregulation and Psychological Well-Being in Elite Sports

The "International Olympic Committee Consensus Statement on Mental Health in Elite Athletes" (IOC - Lausanne, Switzerland, 2018) highlights the necessity of systematically reviewing scientific literature on psychological disorders and symptoms among elite athletes. This symposium underscores the unique psychological vulnerability of athletes who face heightened psychological risk factors due to the ongoing pressure and inherently competitive, intense nature of elite sports. Continuous emotional and physical overload, coupled with the unrelenting drive for excellence, increases the risk of mental health issues. Anxiety, depression, and burnout create a complex and potentially debilitating scenario for elite athletes. The outcomes of these discussions led to a set of recommendations centered around the need for systematic psychological monitoring programs, which include regular assessments of athletes' psychological health. This periodic screening is seen as essential for the early diagnosis of distress symptoms, which, if left unchecked, may compromise both the athlete's health and performance.

This new paradigm proposes a holistic approach to the athlete, where mental well-being topics are equally important to physical training in the emerging narrative. Within this framework, athlete assessment, bio-physiological and psychological monitoring, and field testing of interventions can thus help regulate the sympathetic nervous system, reducing anxiety symptoms and improving athletes' quality of life.

Further studies highlight that the relationship between resilience factors, the negative effects of stress, and the prevention of burnout risk contributes to more excellent emotional stability, improved psychological well-being, and the capacity to perform under pressure (Green et al., 2018; Sarkar et al., 2014).

Moreover, randomized clinical trials emphasize the long-term benefits of mindfulness-based programs (MBPs) on the mental health of elite athletes (Myall et al., 2022).

Such programs integrate embodied awareness approaches that enhance mental resilience, allowing athletes to tolerate competitive stress better. When incorporated into athletes' routines, this approach has proven effective in improving emotional stability and attentional regulation. Numerous studies also underscore the positive impact of mindfulness training on athletes' mental well-being, reducing sport-related anxiety and mental fatigue. In conclusion, these studies emphasize the need to:

- Intervene with greater coherence and precision,
- Increase expertise among coaches and specialized psychologists,
- Adapt these practices more effectively to elite sports.

This paradigm shift not only values psychological resilience as a critical component of performance but also lays the groundwork for a more sustainable, balanced approach to competition in line with the athlete's overall health.

8. Matching and Tailoring: Building a Personalized Intervention Plan

The Matching & Tailoring approach allows for highly personalized interventions tailored to the psychological, social, and environmental characteristics of elite athletes. This method requires a deep analysis of the athlete's insight capabilities and the cultural and relational resources available to them, integrating support from a collaborative multidisciplinary team. Intervention design thus becomes a shared practice between the athlete and staff, enhancing motivation and synergy toward common goals. A significant parallel exists between this approach and the "4P Medicine" model:

- Preventive,
- Participatory,
- Personalized,
- Predictive.

Inherited from systems biology, this paradigm provides an integrated, proactive vision of health, potentially adaptable to sports psychology for forecasting and modulating competitive stress (Hood & Friend, 2011; Martín-

Rodríguez et al., 2024). Sports psychology can benefit from the flexibility and systemic approach of the 4P model, enabling the co-creation of interventions based on biopsychosocial models that predict performance and psychophysical well-being (Bittencourt et al., 2016).

9. The Role of The Sports Psychologist in the Multidisciplinary Team

The inclusion of a sports psychologist in the multidisciplinary intervention team is essential for continuous and accurate monitoring to support both the quality of training and the general health and well-being of athletes in each intervention phase. The IOC symposium (mentioned previously) highlighted the urgency of a systemic, evidence-based approach to managing elite athletes' mental health. The 2018 Consensus Conference emphasized the need to include specialized professionals, such as sports psychologists and psychiatrists, capable of intervening with evidence-based methodologies to improve competitive stress management and emotional regulation, reducing injury risks and negative impacts on overall performance.

Based on these reflections, it is proposed that the specialist in sports psychology may:

1. *Promote* the acquisition of specific, up-to-date skills among sports coaches and colleagues in sports psychology, fostering a complex, integrated perspective on the mind-body relationship and the athlete's global awareness.
2. *Design* interventions that promote a relational alliance between the athlete and practitioner, empathy, and sharing of emotional experiences and body-kinesthetic sensations.
3. *Harmonize* workload programs based on the athlete's psychophysical, familial, and social-contextual characteristics and current sports tasks, preventing psychosomatic disorders, maladaptive stress, and dropout.
4. *Monitor and assess* related psycho-educational variables, fostering medium- and long-term goal adjustment.
5. *Consider*, from a salutogenic and integrated perspective, the athlete's psychological resilience and empowerment processes, promoting

global health and psychophysical well-being with awareness (Cancellara et al., 2022).

10. Specialized Aspects

Using non-invasive medical devices like photoplethysmography (PPG) and biofeedback significantly supports sports psychology specialists in designing targeted protocols for elite athletes. Here is how these tools enhance intervention strategies:

Photoplethysmography (PPG), initially developed for clinical use, is expanding into sports settings to monitor workload and physiological stress. Research by Park et al., (2022) explores how PPG analyzes blood volume variations and is used to monitor variables such as heart rate and heart rate variability, both essential for optimizing athletes' training. These parameters help evaluate the athlete's physiological response in real time, providing essential data for personalized stress and recovery management.

The proposed approach considers integrating tools like PPG into psychophysical performance programs. This approach enables the assessment of the synergy between physical and psychological preparation, which is crucial for maintaining and enhancing performance levels, minimizing risks associated with excessive stress, and improving the overall psychophysical well-being of elite athletes.

Key points are outlined below:

1. Mental State Monitoring and Stress Management

Psychophysical training programs enable the monitoring and management of stress, which can negatively impact performance. Through techniques like biofeedback and mindfulness, athletes develop tools to stay focused under pressure, enhancing their attention, decision-making skills, and emotional regulation during competition.

2. Injury Prevention and Overtraining Avoidance

Elite athletes are constantly at risk of overtraining and subsequent injuries. Structured psychophysical programs help athletes recognize signs of mental and physical fatigue, allowing for the adjustment of training intensity and promoting recovery. Relaxation techniques (such as Jacobson's progressive neuromuscular relaxation, autogenic training, biofeedback, and meditation) and stress management practices also reduce muscle tension, a factor that can contribute to injuries.

3. Photoplethysmography (PPG) for Monitoring Psychophysical State

Photoplethysmography is a valuable tool for monitoring athletes' psychophysical condition. PPG detects changes in blood volume by analyzing reflected light, providing important information on heart rate and heart rate variability (HRV). HRV is a significant indicator for understanding stress levels and autonomic nervous system recovery, allowing coaches to tailor training programs in a personalized way.

4. Recovery Optimization

High-level athletes cannot afford extended recovery periods. Tools like photoplethysmography provide objective data on physiological recovery status, helping to balance training based on the necessary rest. HRV, for instance, is closely linked to vagal tone and thus, to the body's ability to adapt to subsequent efforts. This monitoring type helps identify when an athlete is ready for an intense session or when a less demanding workout is needed.

5. Enhancing Psychological Resilience

Resilience and the ability to maintain control in challenging situations often influence athletic performance. Psychophysical management also includes mental training to foster a winning mindset, supported by visualization techniques, breath control, and body awareness. Measurements such as PPG provide concrete feedback on the effects of these techniques on the body's complex systems. Overall, non-invasive devices such as PPG and biofeedback are helping sports psychology specialists tailor interventions that support elite performance's physical and mental aspects, offering a more holistic approach to optimizing athlete health and performance.

11. Developments, Critical Issues, and Current Challenges

In the rapidly expanding field of wearable PPG sensor applications, clinical studies supported by recent advances in telemedicine and e-health (Nardelli &

Bailón, 2023) play a central role. The significant focus on developing these health monitoring and assessment techniques has fostered a closer integration between healthcare and wellness promotion in sports. In sports, for instance, real-time physical effort tracking and prevention of potentially hazardous conditions offer the opportunity to gather biometric data without interfering with performance, helping maintain high safety levels during training (Nardelli & Bailón, 2023). These core attributes are factors that sports psychologists might consider both in specialized training programs and in multidisciplinary collaboration frameworks.

Recent research explores the integration of photoplethysmography (PPG) in psychophysical performance monitoring programs for elite athletes, covering developments, practical applications, and current challenges. PPG and its variants, such as remote PPG (rPPG), are valuable in integrating psychophysical monitoring programs for elite athletes. PPG allows for accurate assessment of critical parameters like heart rate variability, oxygen saturation, and, in some instances, stress levels and sleep quality. Literature reviews highlight that PPG, often combined with machine learning methods, has significant potential to enhance real-time biometric and metabolic parameter detection. Supported by wearable devices, this technology offers a non-invasive, cost-effective solution, providing real-time data seamlessly integrated into training programs.

Considering PPG's applications, primarily in healthcare for health monitoring, it is essential to account for the complexity of analyzing nonlinear data, which also brings attention to dynamic properties such as "determinism," "predictability factors," and "entropy" in experimental settings.

However, several technical disadvantages remain, such as minimizing errors caused by movement and skin pigmentation variations and improving result accuracy. These technical and scientific challenges are being addressed through the implementation of advanced algorithms and the use of multispectral signals (Kim & Baek, 2023). In this context, numerous challenges related to PPG device design and various data processing techniques persist, fueling the increasing interest in new signal processing methodologies.

Conclusion

The introduction of psychophysical performance programs, supported by advanced tools like PPG, provides professional and elite athletes a competitive advantage. This integrated approach maximizes training efficacy, reduces injury and burnout risks, and optimizes psychophysical recovery. The IOC symposium anticipated a paradigm shift centered on a complex, integrated vision of the athlete, promoting a holistic perspective that unites physical, psychological, and social dimensions. This biopsychosocial view acknowledges that elite athletes operate in a dynamic equilibrium, where psychological components and managing social and environmental relationships are crucial for resilience and emotional stability.

Our intervention vision in sports psychology aligns with the IOC's guidelines and recommendations to protect the health and well-being of athletes (even young adolescents facing vulnerable life stages) and facilitate a safe and supportive environment for international sports competitions (Bergeron et al., 2024).

Finally, our study shows the protocol's effectiveness in all athletes. In particular, the use of PPG in combination with biofeedback provides valuable insights into the athletes' state of adaptation. It facilitates workload management, recovery strategies, and mental preparation activities tailored to each athlete and team member.

References

Aubert, A. E., Seps, B., & Beckers, F. (2003). Heart rate variability in athletes. *Sports Medicine*, 33(12), 889–919. https://doi.org/10.2165/00007256-200333120-00003.

Bellenger, C. R., Fuller, J. T., Thomson, R. L., Davison, K., Robertson, E. Y., & Buckley, J. D. (2016). Monitoring athletic training status through autonomic heart rate regulation: A systematic review and meta-analysis. *Sports Medicine*, 46(10), 1461–1486. https://doi.org/10.1007/s40279-016-0484-2.

Bergeron, M. F., Côté, J., Cumming, S. P., Purcell, R., Armstrong, N., Basilico, L., Burrows, K., Charrin, J. B., Felix, A., Groesswang, H., Iwasaki, Y., Kocher, M. S., Martowicz, M., McConnell, K., Moran, J., Holm Moseid, C., Mountjoy, M., Soligard, T., Tetelbaum, E., Thiel, A., Vertommen T., Viseras G., Budgett R., Engebretsen L., & Erdener, U. (2024). IOC consensus statement on elite youth athletes competing at the Olympic Games: essentials to a healthy, safe, sustainable paradigm. *British journal of sports medicine*, 58(17), 946–965. https://doi.org/10.1136/bjsports-2024-108186.

Bittencourt, N. F. N., Meeuwisse, W. H., Mendonça, L. D., Nettel-Aguirre, A., Ocarino, J. M., & Fonseca, S. T. (2016). Complex systems approach for sports injuries: moving from risk factor identification to injury pattern recognition—narrative review and new concept. *British Journal of Sports Medicine,* 6; 50:1309-1314. http://dx.doi.org/10.1136/bjsports-2015-095850.

Blumenstein, B., Orbach, I., Aaron, R., Weinstein, Y., & Moran, D. (2017). The effect of biofeedback training on anxiety in physical education Teachers short title: biofeedback training and PE teacher's anxiety. MOJ Yoga. Physical Ther; 2(3):74-77. DOI: 10.15406/mojypt.2017.02.00022.

Cancellara, L., & Ballocco, S. (2022). *Chaos and Complexity Letters*, suppl. Special Issue: *Complexity Science in Human Change*; Hauppauge Vol. 16, 3: 171-185. ISSN: 1556-3995, Nova Science Publisher.

Clarsen, B., Myklebust, G., & Bahr, R. (2013). Development and validation of a new method for the registration of overuse injuries in sports injury epidemiology: The Oslo Sports Trauma Research Centre (OSTRC) overuse injury questionnaire. *British Journal of Sports Medicine*, 47(8), 495-502. http://dx.doi.org/10.1136/bjsports-2012-091524.

Davids, K., Hristovski, R., Araújo, D., Balague Serre, N., Button, C., & Passos, P. (2014). *Complex systems in sport*. London: Routledge.

Fletcher, D., & Sarkar, M. (2012). *A grounded theory of psychological resilience in Olympic champions. Psychology of Sport and Exercise, 13*(5), 669–678. doi: 10.1016/j.psychsport.2012.04.007.

Giges, B., & Petitpas, A. (2020). The Sport Psychologist's Role in Addressing Mental Health and Performance Issues in Sports. *The Sport Psychologist*, 34(4), 281-289.

Green, J., Rees T., Peters, K., Sarkar, M., & Haslam, S. A. (2018). Resolving Not to Quit: Evidence That Salient Group Memberships Increase Resilience in a Sensorimotor Task. *Front. Psychol.* 9:2579. https://doi.org/10.3389/fpsyg.2018.02579.

Gustafsson, H., Skoog, T., Davis, P., Kenttä, G., & Haberl, P. (2015). Mindfulness and its relationship with perceived stress, affect, and burnout in elite junior athletes. *Journal of Clinical Sport Psychology, 9*(3), 263–281. https://doi.org/10.1123/jcsp.2014-0051.

Henriksen, K., Schinke, R., Moesch, K., McCann, S., Parham, WD., Larsen, C. H., & Terry, P. (2020). Consensus statement on improving the mental health of high-performance athletes. *International Journal of Sport and Exercise Psychology*, 18(5), 553-560. https://doi.org/10.1080/1612197X.2019.1570473.

Hood, L., & Friend, S. H. (2011). Predictive, personalized, preventive, participatory (P4) cancer medicine. *Nature Reviews Clinical Oncology*, 8(3), 184–187.

Kim, K. B., & Baek, H. J. (2023). Photoplethysmography in Wearable Devices: A Comprehensive Review of Technological Advances, Current Challenges, and Future Directions. *Electronics* 12, 2923. https:// doi.org/10.3390/electronics12132923.

Martín-Rodríguez, A., Gostian-Ropotin, L. A., Beltrán-Velasco, A. I., Belando-Pedreño, N., Simón, J. A., López-Mora, C., Navarro-Jiménez, E., Tornero-Aguilera, J. F., & Clemente-Suárez, V. J. (2024). Sporting Mind: The Interplay of Physical Activity and Psychological Health. *Sports (Basel)*. 12(1):37. doi: 10.3390/sports12010037. PMID: 38275986; PMCID: PMC10819297.

Myall, K., Montero-Marin, J., Gorczynski, P., Kajee, N., Syed Sheriff, R., Bernard, R., Harriss, E., & Kuyken W. (2023). Effect of mindfulness-based programmes on elite athlete mental health: a systematic review and meta-analysis. *Br. J. Sports Med.* 2023 Jan; 57(2):99-108. doi: 10.1136/bjsports-2022-105596. Epub 2022 Oct 12. PMID: 36223914.

Nardelli, M., & Bailón, R. (2023). Advances in Wearable Photoplethysmography Applications in Health Monitoring. *Sensors (Basel, Switzerland)*, *23*(16), 7064. https://doi.org/10.3390/s23167064.

Park, J., Seok, H. S., Kim, S.-S., & Shin, H. (2022). *Photoplethysmogram Analysis and Applications: An Integrative Review. *Frontiers in Physiology* https://doi.org/10.3389/fphys.2021.808451.

Reardon, C. L., Hainline, B., Aron, C. M., Baron, D., Baum, A. L., Bindra, A., Budgett, R., Campriani, N., Castaldelli-Maia, J. M., Currie, A., Derevensky, J. L., Glick, I. D., Gorczynski, P., Gouttebarge, V., Grandner, M. A., Han, D. H., Mcduff, D., Mountjoy, M., Polat, A., Rosemary Purcell R., Putukian, Rice, S., Sills, A., Stull, T., Swartz, L., Jing Zhu, & L., Engebretsen, L. (2019). Mental health in elite athletes: International Olympic Committee consensus statement. *British Journal of Sports Medicine*, 53(11), 667–699. https://doi.org/10.1136/bjsports-2019-100715.

Rees, C. S., Breen, L. J., Cusack, L. & Hegney, D. (2015). Understanding individual resilience in the workplace: the international collaboration of workforce resilience model. *Front. Psychol.* 6:73. doi: 10.3389/fpsyg.2015.00073.

Sarkar, M., & Fletcher, D. (2014). Psychological resilience in sport performers: a review of stressors and protective factors. *J Sports Sci.* 32(15):1419-34. doi: 10.1080/02640414.2014.901551. Epub 2014 Apr 9. PMID: 24716648.

Schinke, R., Wylleman, P., Henriksen, K., Si, G., Wagstaff, C. R. D., Zhang, L., Tshepang, T., Noce, F., & Li, Y. (2023). International Society of Sport Psychology position stand: scientist practitioners. *International Journal of Sport and Exercise Psychology*, 22(1), 1-23. https://doi.org/10.1080/1612197X.2023.2174681.

Stecchi, F., & Leone, S. (2022). Explorative Counselling Framework. The Caos and Complexity of the Athlete's Emotional Experiences: The Insynergy Sport Fast Emotional Insight (ISFEI) Game Cards. *Chaos and Complexity Letters*, Volume 16, 3, 159–169. ISSN: 1556–3995, Nova Science Publisher.

Tossici, G., Zurloni, V. & Nitri, A. (2024). Stress and sport performance: a PNEI multidisciplinary approach. *Front. Psychol.* 15:1358771. https://doi.org/10.3389/fpsyg.2024.1358771.

Chapter 10

Entrepreneurship as a Distinct and Indistinct Variable: The Complex Critique of Conventional Economic Theory

Gabriele Serafini*
Niccolò Cusano University, Rome, Italy

Abstract

Complexity economics criticises conventional economic theory based on two results that undoubtedly characterise it: general economic equilibrium and the representation of the overall economic system using a structure analogous to that of a microeconomic system. However, the two criticisms do not affect the methodological foundation of conventional economic theory. It is founded on determining the value of goods at the moment of their exchange on the market, and an effective criticism must, therefore, be directed towards the market price through an investigation of the elements of the price and the places of its determination. The distinction of entrepreneurship from business management and its indistinguishability in the value of each product on the market constitutes precisely the elements that can be used to highlight the limits of conventional economic theory and to delineate the complexity of economic value and, thus, of economics as a science, far beyond what is considered correctly established through the market price.

Keywords: complexity economics, Neoclassical economics critique, Theory of value
JEL: B25; B41; P42

* Corresponding Author's Email: gaserafin@yahoo.it

In: Complexity and Fractality in Human Change and Beyond
Editor: Nicoletta Sala
ISBN: 979-8-89530-575-1

1. Introduction: The Two Critiques by Complexity Economics

Scholars who can be traced back to the line of research of complexity economics criticise the results of conventional economic theory - often also referred to as standard or orthodox (Blaug, 1970 [1968], p. 10) (Landreth & Colander, 1996, p. 353) – concerning two specific areas: the general economic equilibrium and the isomorphism that would characterise macroeconomics in relation to microeconomics (Arthur, 1999, p. 107; Arthur, 2021, p. 143; Dosi & Roventini, 2019, p. 2; Fagiolo, 2012; Caiani, et al., 2016). This chapter will first state the essence of the critiques and then the reasons why they cannot be considered relevant. The conclusions will explain why economic value is a complex variable and how this consideration qualifies economic science as a complex science.

As W. Brian Arthur writes about the first critique, "Complexity economics sees the economy as in motion, perpetually 'computing' itself - perpetually constructing itself anew. Where equilibrium economics emphasises order, determinacy, deduction, and stasis, complexity economics emphasises contingency, indeterminacy, sense-making, and openness to change" (Arthur, 2013, p. 1). This clarifies how proponents of a complex approach to economic theory qualify the economic system as evolutionary and, in general, outside of equilibrium conditions.

Concerning the second critique, on the other hand, the object of attention is the *Dynamic Stochastic General Equilibrium* - DSGE - models (Christiano et al., 2018, p. 113; Torres, 2016, p. 97). These models constitute a description of the behaviour of overall economic systems, not of their parts. They, as macroeconomics, are distantly derived from the work of J. M. Keynes (1978 [1936]) and are thus aimed at modelling overall economic systems rather than portions of them. The qualifying feature of such models consists in their description based on models of economic agents, representative of the actually existing agents, who would act by correctly estimating, albeit on average, the future values of economic variables (Dosi & Roventini, 2019, p. 9; McDonald, 2022, p. 174; Romer, 2019, p. 309 ff.) (Blanchard, 2018, p. 23 ff.). In this case, the criticism centres on reducing the heterogeneity characterising the behaviour of the various agents present in economic systems to the behaviour of a single agent, representative of the behaviour of all, making the macroeconomic level analogous to a microeconomic one. This results in a loss of complexity, the value of which would have recently been reinforced by the development of real economies compared to the estimates of the DSGE forecast models (Saraceno, 2021).

Indeed, criticism in recent years has focused on the fact that representative economic agents, and thus the models themselves, should have predicted the Great Recession that has characterised the economic system since 2007-2008. However, they did not do so (Lucarelli, 2021). The forecast error is not attributed to an inaccurate estimate but to the essence of the DSGE models, which neglect the heterogeneity of the real economic agents and, through this, come to deny *a priori* the possibility of a crisis. The absence of heterogeneity and the attribution of a probability of occurrence for a crisis event are thus summarised in the inability to describe economic systems correctly. The reason for this is traced back to the actual complexity of the behaviour of the various economic agents, which leads to a qualitative difference between the description of micro-level behaviour and that of macro-level behaviour of the overall system. The criticism is therefore directed at the isomorphism between the two levels of analysis, micro and macro, compared to what should be correctly detected in reality, with its qualitative differences between a micro and a macro level of economic systems.

2. The Irrelevance of the Two Criticisms

If these are the two criticisms, about the first, it should be noted that the economic literature shows that general economic equilibrium is not an essential outcome for the validity of that methodological approach (Serafini, 2021, p. 7 ff.). The attainment of general economic equilibrium constitutes a strand of study within conventional theory, and for this reason alone, its results could not invalidate its assumptions. In any case, general equilibrium consists, according to the neoclassical literature itself, of an ideal result that any economic system could achieve if all market forces, favourable and opposing, characterising supply and demand, were allowed to act within a logical time horizon (Walras, 2006 [1874], p. 322; Pareto, 2006 [1906], p. 113). In the neoclassical economics framework, time is necessarily logical and not historical, as emphasised by all scholars on the subject, since the description of the achievement of a general economic equilibrium requires the instantaneousness and simultaneity of all production and exchange processes (Agnati, 2001, p. 362; Graziani, 1993, p. 471). Literature has also been spent extensively on criticism of the results of this line of research, as well as on the realism of the assumptions underlying it. In order for the criticism of complexity economics to be effective, it is, therefore, necessary to consider that it is precisely from within the neoclassical approach, which constitutes the foundation of conventional economic theory, that fundamental criticism of the relevance of these results and its assumptions has always been voiced. In particular, the Austrian School, starting with the work of the early theorists of the neoclassical conception itself, immediately set up its theory through a critique of the rationality of agents and the perfect information available in the markets (Menger, 1985 [1883], p. 41 ff) (Menger, 1976 [1871], p. 234), which should characterise economic systems for general economic equilibrium to be conceivable. These criticisms, therefore, from the beginnings of the neoclassical approach and also through other fundamental theorists, both from the origins (Marshall, 2006 [1890], p. 97 ff.) and more recent ones in this field (von Hayek, 1997 [1949], p. 49) (1994 [1982], p. 269) (von Mises, 2003 [1933], 168) as well as through other eminent economists critical of general economic equilibrium (Robinson, 1991, p. 314), arrive at the impossibility of basing the value of each economic quantity on the value they should theoretically have under an unrealisable economic condition. The criticism, however, is internal to the conventional conception since human action is not considered to be characterised by perfect knowledge of reality and never can be. This leads these authors to dispute that the values of goods, which would correspond to that ideal situation, could ever be the values that actually exist. Market prices would then necessarily differ from the prices corresponding to those unrealistic conditions. The contestation of the results of the value of goods as determined using an unrealistic general economic equilibrium, precisely because it refers to the unrealistic conditions, however, only affects the ideality of that result, but not the correctness of the market process that would underlie the determination of the value of goods, which is represented by their price. For this reason, the difference between adherence to the hypothesis of the ideality of the general economic equilibrium and, on the contrary, opposition to this ideality, because it is considered unrealistic, does not invalidate the common belonging of these two visions to the same conventional economic conception.

However, the criticism of the general economic equilibrium conception is considered adequate by the theorists of complexity economics, who end up neglecting the fact that the very theorists of methodological individualism who support it, including Friedrich von Hayek, often cited to support the critique by complexity economics, have an approach that is actually and theoretically internal to the same approach as conventional economic theory.

Therefore, this theoretical reference can be considered as a mistake made by complexity economics. An error that confirms that the result of the general economic equilibrium is not, in reality, essential for the determination of the value of goods, whereas, as will be seen, the reference of all strands of thought within the neoclassical conception to market exchange and the determination of the value of goods coinciding with the price they receive at the moment of exchange, would be. About the second criticism, however, it is necessary to consider that the original macroeconomic approach, as expressed by John Maynard Keynes in his '*General Theory*' (Keynes, 1978 [1936], p. 185), considered that the whole is different from the parts. This implied that, instead of proceeding from a micro-foundation of macro-economics, the Keynesian origin approach proceeded from a sort of macro-foundation of economic behaviour (Graziani, 1992, p. 13; Demaria, 1961), and thus that there was a difference between the micro and macro level of economic relations, in terms of descriptive variables. It is precisely the conception expressed by Keynes that indicates how, at the micro level, entrepreneurs invest according to their expectations of earnings, but at the macro level, consumers in a progressively richer society spend a smaller portion of their income (Keynes, 1978 [1936], p. 189). For this reason, at the macro level, the portion of income spent on consumption tends to decrease over time as the growth of wealth-produced proceeds; consequently, entrepreneurs' investment also tends to grow more slowly, if at all, because it is linked to expectations of reduced sales, in the face of productivity growth that gradually makes it possible to obtain the same amount of product using fewer resources. It should be noted that a variation in productivity that implies a possibility to produce with less use of resources, necessarily also changes the quality of production and not only its quantity. This is because if each product is produced using a different production process, it necessarily follows that it is qualitatively different. Due to this work's limitations, it is impossible to dwell on this aspect. However, it is also relevant to understand how economic value, as pointed out in the concluding section, is a complex quantity.

The change in productivity and the consequent use of fewer resources explains how, if needs do not increase to keep up with productivity, an increase in wealth obtained at the micro level determines the re- emergence of poverty at the macro level, and thus how a kind of macro-foundation of microeconomic behaviour represents the different characteristics of the two levels of analysis.

Other later Keynesian approaches, such as that expressed by Franco Modigliani (1944, p. 45), for example, make it clear how the behaviour of a set of subjects individually determines, at the macro level, a particular prevalence of one type of behaviour in the overall system rather than another. In this way, conventional economic literature of the 20th century also portrayed how the heterogeneity of behaviour can determine different characteristics at the macro level in relation to the prevalence of certain types of behaviour rather than others. That is, in this case, the different macroeconomic characteristics depend on the prevalence of certain microeconomic behaviours rather than others, which are, in any case, heterogeneous.

Before Keynes, however, again concerning the differences in the qualification of analysis between a micro and a macro level, reference can be made to the work of Karl Marx. In this case, it is well known the conception according to which the differences between the behaviour of each capitalist are emphasised, who at the micro level would fight in the competitive system to expel the other capitalists from the market by behaving like a false brother, while at the macro level, he would fight together with the other capitalists, uniting with them in the class struggle against the workers (1989 [1894], p. 242). The difference and coexistence of the two analysis levels have long been present in the literature, although not necessarily enunciated and made evident as in complexity economics.

It is well then to specify how the critique elaborated by complexity economics is not addressed to the Keynesian theory of its origins, nor to the later or the preceding economic theories, but rather properly to the DSGE models and to the related recent literature that simplifies reality by compressing the heterogeneity of consumer characteristics into a single behaviour of a representative consumer. This criticism is therefore centred on the current literature but, precisely insofar as it relates to the aspects of the heterogeneity of behaviour and the different characteristics that must describe the microeconomic and macroeconomic levels of analysis, it does not constitute a novelty in the panorama of the economic literature and, above all, as pointed out earlier, does not contradict the methodological approach that underlies the conventional neoclassical conception.

3. Entrepreneurship and the Elements That Makeup Products on the Market

To be effective, economic criticism must instead be directed at the analytical heart of conventional theory, namely, the price determined when supply and demand meet. The neoclassical conception is founded on determining the value of economic quantities at the moment of market exchange due to the interaction of supply and demand (Landreth & Colander, 1996, p. 482).

Therefore, as long as this determination is not questioned, we consider non-essential points of economic theory to be attacked. It is, therefore, necessary to conceive of a critique of the market price as a signal and indicator of the relative values of economic quantities, which makes it clear that it is not determined in the market exchange by the forces of supply and demand. In particular, in an economy that is not one of pure exchange but also includes production, the market price of products must be the object of criticism.

Production constitutes, according to conventional theory, the process by which production factors and intermediate goods are combined to transform initial resources into final resources (Pareto, 2006 [1906], p. 112) (Walras, 2006 [1874], p. 324). The initiative of the enterprise carries out this process, understood as that entity, be it public or private, which is responsible for the initial choices to initiate a production process. The management of the production and exchange processes can then be delegated to workers who manage the enterprises or to any capital device that can perform the assigned tasks (Walras, 2006 [1874], p. 343). This means that one of the essential elements for the start-up of the production process is entrepreneurship, understood broadly as that element that induces the start-up of productive activity. Clearly and historically, this activity was first predominantly carried out by individual private actors, whereas it is now also carried out by public sector actors, employing an activity that has recently been recognised as having a first-rate entrepreneurial role (Mazzucato, 2014).

Entrepreneurship, therefore, constitutes one of the essential elements of production that represents, according to conventional economic theorists, a 'fourth character' (Walras, 2006 [1874], p. 319) treading the stage of the modern economic system. As such, her/his activity must be remunerated, and her/his remuneration is called profit. This remuneration is therefore considered among the income receiving a share of the net price obtained on the market through the sale of the product. If an activity is not remunerated, on the other hand, in an economic system dominated by production for market exchange, the activity exercised tends not to be able to be reproduced. The elements that receive remuneration from the market thus include the inputs used and the intermediate goods fed into the production process.

In this view, it is necessary to turn the attention to entrepreneurship because this activity constitutes the only one, among those involved in the production process, that is not traded on the market. The other elements that determine the realisation of products, such as intermediate goods and the services of the various factors of production, on the other hand, have a market price; it is, therefore, necessary to carefully qualify entrepreneurship because the fact that, at the same time, entrepreneurship constitutes a fundamental element in the realisation of products, is remunerated through the allocation of a part of the market price, and does not constitute a variable exchanged on the market, constitutes precisely the weak point of the conventional theory of value. This element needs to be better explained, beginning by clarifying that entrepreneurship has two possible determinations in conventional economic theory. According to the version of the theory that does not consider adequate to represent the economic system using the possibility of reaching a position of general economic equilibrium (Marshall, 2006 [1890], p. 816) (von Mises, 2000 [1978], p. 96) (von Hayek, 1997 [1949], p. 45) entrepreneurship is a capacity that differs from individual to individual, which may or may not be possessed by a subject. In this sense, a particular activity could be considered a service sold by each monopolist owner. As we shall see in the next paragraph, this is incorrect because, since entrepreneurship cannot be distinguished from the product it contributes to create, it is not an autonomous market quantity concerning the outcome of production, and its remuneration, profit, cannot, therefore, be considered a market price.

According to the version of economic theory that, on the other hand, contemplates general economic equilibrium, entrepreneurship constitutes a quantity analogous to business management activities (Walras, 2006 [1874], p. 323) (Pareto, 2006 [1906], p. 166). In this case, profit is regarded as the remuneration of an activity analogous to the business management activity bought and sold on the market. For this reason, profit is set at a certain level in the market, given by the price that the business management has in the market. Suppose the profit is rationally considered as a cost incurred by the business to produce. In that case, as conventional economic theory affirms, businesses make neither profits nor losses under competitive equilibrium conditions.

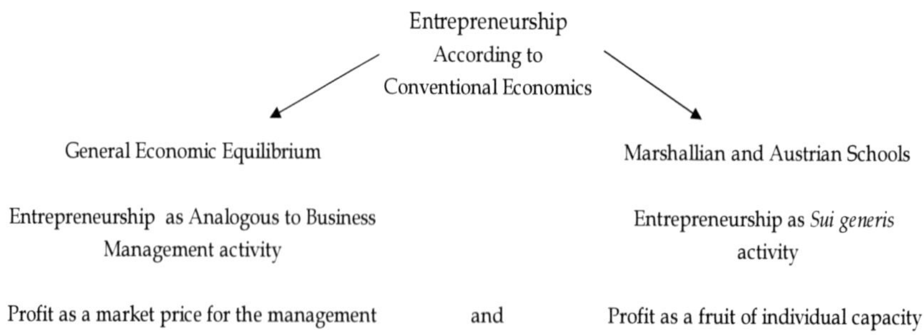

Scheme 1.

These two alternative ways of qualifying the economic system characterise conventional economics and, it may be argued, both constitute evidence of the insurmountable limitation that the market price mechanism encounters if it is to be considered the mechanism capable of representing how the exchange value of economic quantities is formed. It is necessary to clarify the origin of this aspect due to entrepreneurship qualification.

4. The Dual Qualification of Entrepreneurship as a Distinct and Indistinct Quantity

First of all, it must be made clear that, in both cases, profit, as remuneration for entrepreneurship, cannot be considered a quantity directly established on the market because entrepreneurship, according to conventional economics, is not something that is traded on the market. The conventional approach, therefore, states that profit constitutes the remuneration of entrepreneurial activity, although it can be understood in two ways. It can be regarded as an activity *sui generis*, or analogous to a business management activity.

In the first case, entrepreneurial activity is considered the fruit of individual capacities, which would, therefore, allow the entrepreneur to earn a remuneration and profit that would have no analogue on the market. Her/his activity should then be able to be distinguished from others, and her/his remuneration, profit, should be able to be obtained autonomously concerning the sale of the product that she/he has helped to produce through the provision of the service and its application to the initial resources.

In the second case, on the other hand, entrepreneurial activity is considered analogous to an activity bought and sold on the market, which would receive a market price for its services: the management salary. In this case, profit would not be distinguished from the management salary, except by the name commonly given to it. Although it does not directly receive a market price, in fact, its amount would be established at a similar level to the market price of the management salary precisely because it is the remuneration of an activity corresponding to that.

These two definitions by the conventional economics approach must be overcome, both because of the differences between entrepreneurial activity and business management and because of the impossibility of distinguishing the contribution of entrepreneurship in the qualification of a product with respect to the other characteristics that can make it recognisable on the market.

Therefore, starting from the second case, it is necessary to ask how entrepreneurial activity can be qualified as analogous to business management. It is clear that the management of an enterprise concerns activities undertaken, precisely, to manage an enterprise; not to start one. That is to say, when one starts a business, that is, one makes the choice to start a business, there is no managerial salary that can remunerate this activity because it is not a managerial activity. Entrepreneurial activity is, in this sense, clearly distinct from other activities and, in particular, from business management. This is also argued, to some extent, by an eminent liberal economist, Mark Blaug, who states that "a businessman need not be a 'capitalist' or 'manager' but he must be a decision-maker, whether he liked it or not. It is this function and this function alone that deserves the title of 'entrepreneurship'" (Blaug, 1970 [1968], p. 76).

Even a business manager makes a choice; however, the entrepreneur decides whether or not to start a business. The literature, however, already distinguishes between the two activities and holds that entrepreneurship is not an activity that can be directly bought and sold on the market. For this reason, one cannot consider profit as a market-established quantity, of course, but it is also incorrect to consider profit as qualitatively similar to a market-established and quantitatively equivalent remuneration. The part of profit that remunerates a managerial activity, that of business management, is indeed corresponding to a management salary, but the part that is earned by the entrepreneur as such, i.e., for an activity that does not correspond to a managerial activity, cannot be considered analogous to a wage. This implies that profit, understood as the part of the product price that remunerates entrepreneurial activity, does not correspond to the remuneration of an activity traded on the market. Therefore, the market price cannot be regarded as consisting of a set of market prices of activities, nor can it be regarded as divided, i.e., as distributed income, between activities remunerated according to their respective market prices. This means that at least one of the elements characterising the modern and contemporary economy and the very qualification of market products does not have a market price. This implies that the value of variables exchanged on the market is generally not determined, even if the exchange of commodities occurs on the market. In the first case, on the other hand, where entrepreneurship is considered a particular activity, the result of individual skills possessed by the entrepreneur, the distinction from business management activity is already evidently recognised. Conventional economic literature, in such a conception, while continuing to hold that entrepreneurship is not bought and sold, believes that the contribution of different entrepreneurship has a market recognition through the quality of the products and the relative price they obtain on the market (Marshall, 2006 [1890], p. 819).

In this case, it should be considered that if two products differed only in the two different entrepreneurship that determined them, they should only be recognised as different because of the entrepreneurship and not because of the other characteristics, for which the two products should not be distinguishable.

A logical problem arises here. If two products are the same in all characteristics except for entrepreneurship, it is claimed that entrepreneurship would be a characteristic clearly distinguishable from the other characteristics of a product.

However, this is not stated in the literature precisely because entrepreneurship is not a characteristic of a product but is instead regarded as the activity corresponding to the combination of initial resources for their transformation into final resources.

Two different entrepreneurships would create two different products. If the two products are different, this means that they have different characteristics insofar as they are perceived as satisfying two different needs or as satisfying the same need differently. Therefore, they are different because of their own perceived characteristics on the market, and for this reason, the two products are not the same. They are qualified as different products and not, contrary to the conventional theoretical approach, as equal products with different entrepreneurship.

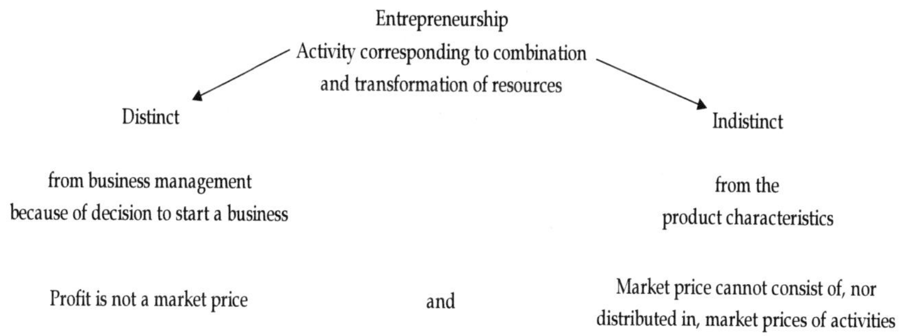

Scheme 2.

Therefore, it is not correct to consider profit as a quantity established on the market, of course, but insofar as one cannot even consider entrepreneurship as a quantity distinguishable from the other elements - which alone would be the only difference between two products on the market - profit is not, even in this case, a market price, even indirectly, but precisely because it cannot be considered remuneration for an indistinguishable element in the product, unlike the other identifiable and marketable elements of the product.

Conclusion: Economic Science and Value as a Complex Magnitude

From the above, it can be concluded that entrepreneurship is a distinct variable from business management and, at the same time, is an indistinct variable from the characteristics of products on the market. This means that at least one of the fundamental elements for the very existence of products is not a market quantity and that its value cannot be found on the market, neither directly, therefore, nor indirectly. The value of economic quantities does not coincide with the market price, and determining the value of economic quantities does not depend only on the moment of exchange. Moreover, the production process of obtaining the product depends on the production process, which includes the initial decisions on the production organisation, the management decisions on the process, and the final decisions on the allocation of the resources obtained. Therefore, it is consistent to assume that value does not depend only on the moment of exchange on the market but on the social conditions in which the economic process occurs. This implies that the value of economic quantities cannot be represented by the market price but rather has to be represented by the value attributions made during the economic process in the relevant social context.

The reason why economic science can thus be considered a complex science goes far beyond what apparently, but not crucially, can be established by a critique of the general economic equilibrium or isomorphism that would characterise macro-level versus micro-level economic systems. The qualification of entrepreneurship, and therefore of economic values as quantities not established by the market, characterises values as complex quantities insofar as they are not determined by an immediate relationship between certain market premises and the result that can be logically deduced from these premises in the market. The value of economic quantities, and thus their price, depends on and emerges from the unfolding of the overall process of the economic system, i.e., the production, exchange, distribution, investment and reproduction phase. It thus encompasses the stages of the process that occur in historical time, during which economic subjects make decisions in relation to other subjects and within a broader social context. The overcoming of the market price as a reference for the determination of value must therefore be realised through the consideration of this process of determination that is more complex than the market exchange process alone, in which the qualifications of the economic subjects involved also play a role, as advocated by complexity economics; however, value will not qualify as a complex element and, in this sense, economic science will not qualify as a complex science, not insofar as it is deterministic rather than probabilistic, but as long as the data used continue to be extracted from the exchange markets.

Overcoming reference to the market is thus the central element on which the pricing of economic quantities must be based, and thus, economic science itself as a science that takes note of the complexity of economic phenomena. The latter, i.e., the phenomena that should constitute the object of economic science, are therefore not the market phenomena alone, with heterogeneous subjects acting in a historical time, but are the different processes of value determination that in the market obtain a certain dimensioning in the form of the price recorded by companies, but are determined by other conditions, to be traced as much in times as in spaces that are not only those of the market. Only by studying these phenomena will it be possible, at the same time, to develop an adequate theory of value, which is therefore not based on observation and the computational elements permitted by the current capacities for calculating quantities valued in the markets but which considers the ways of relating between subjects and between subjects and the context of reference. Hence, therefore, considers the context and historical interactions, the choices that economic subjects make, both individually and as parts of the context to which they feel they refer, and how they plan and carry out the activities to be performed, which economic subjects have always carried out to satisfy needs, perceived as such.

References

Agnati, A. (2001). Critica dei massimi sistemi dell'economia politica. Padova: CEDAM.

Arthur, W. B. (1999). Complexity and the Economy. *Science*, Volume 284, 107–109.

Arthur, W. B. (2013). C*omplexity Economics: A Different Framework for Economic Thought.* SFI Working paper.

Arthur, W. B. (2021). Foundations of complexity economics. *Nature Physics*, pp. 136–145.

Blanchard, O. (2018). On the future of macroeconomic models. *Oxford Review of Economic Policy*, 43–54.

Blaug, M. (1970) [1968]. *Storia e critica della teoria economica.* Torino: Boringhieri.

Caiani, A. et al., (2016). Agent Based-Stock Flow Consistent Macroeconomics: Towards a Benchmark Model. *Journal of Economic Dynamics & Control,* Issue 69, 375–408.

Christiano, L. J., Eichenbaum, M. S., & Trabandt, M. (2018). On DSGE Models. *Journal of Economic Perspectives*, 113–140.

Demaria, G. (1961). *Macro e microeconomia.* Roma: Istituto dell'Enciclopedia Italiana Treccani.

Dosi, G. & Roventini, A. (2019). More is different ... and complex!: The case for agent-based macroeconomics. *LEM Working Paper Series,* Issue 01, 1–37.

Fagiolo, G. (2012). *Complessità economica.* Roma: Istituto dell'Enciclopedia Italiana Treccani.

Graziani, A. (1992). *Teoria economica. Macroeconomia.* Napoli: Edizioni Scientifiche Italiane.

Graziani, A. (1993). *Teoria economica. Prezzi e ditribuzione.* Napoli: Edizioni Scientifiche Italiane.

Keynes, J. M. (1978). [1936]. *Teoria generale dell'occupazione, dell'interesse e della moneta*. Torino: UTET.

Landreth, H., & Colander, D. C. (1996). *Storia del pensiero economico*. Bologna: il Mulino.

Lucarelli, S. (2021). L'economia ha bisogno della complessità. Nella factory degli economisti underground. *Social and political notes* [Online].

Marshall, A., (2006). [1890]. *Principi di economia.* Milano: Milano Finanza.

Marx, K. (1989). [1894]. Il capitale. *Critica dell'economia politica.* Libro III. Roma: Editori riuniti.

Mazzucato, M. (2014). *Lo Stato innovatore*. Bari-Roma: Editori Laterza.

McDonald, J. F. (2022). *Rethinking Macroeconomics*. London and New York: Routledge.

Menger, C. (1976) [1871]. *Principi di economia politica.* Torino: UTET.

Menger, C. (1985) [1883]. *Investigation into the method of the social sciences with special reference to economics.* New York: New York University Press.

Modigliani, F. (1944). Liquidity Preference and the Theory of Interest and Money. *Econometrica*, 12(1), 45–88.

Pareto, V. (2006) [1906]. *Manuale di economia politica.* Milano: Egea.

Robinson, J. V. (1991). *Occupazione distribuzione e crescita. B*ologna: il Mulino.

Romer, D. (2019). *Advanced Macroeconomics*. New York: McGraw-Hill.

Saraceno, F. (2021). Per un'economia della complessità. In: *Il mercato rende liberi*. Roma: LUISS University Press.

Serafini, G. (2021). Complexity Economics and Neoclassical Economics. A Critique. *Chaos and Complexity Letters,* 15(1), pp. 7–16.

Torres, J. L. (2016). *Introduction to Dynamic Macroeconomic General Equilibrium Models.* Wilmington: Vernon Press.

von Hayek, F. A. (1994) [1982]. *Legge, legislazione e libertà.* Milano: Saggiatore.

von Hayek, F. A. (1997) [1949]. *Individualismo: quello vero e quello falso.* Soveria Mannelli: Rubbettino.

von Mises, L. (2003) [1933]. *Epistemological problems of economics.* Auburn, Alabama: Ludwig von Mises Institute.

von Mises, L. (2000) [1978]. Profit and Loss. In: R. Swedberg, *a cura di Entrepreneurship.* Oxford: Oxford University Press, 89–109.

Walras, L. (2006) [1874]. *Elementi di economia politica pura.* Milano: Milano Finanza Editore.

Chapter 11

The Application of the Log Decrement and the Log Increment to Identify Parameters of an Oscillation

Javier Montenegro Joo*
Facultad de Física, Universidad Nacional Mayor de San Marcos, UNMSM, Lima, Perú
VirtualDynamics: Science and Engineering Virtual Labs, Lima, Peru

Abstract

Based on the technique of logarithmic decrement to identify the parameters of a damped oscillation, an educational virtual lab to determine the parameters of a stimulated oscillation has also been developed. At the experiment's opening, the virtual lab's operator enters the input data of the oscillation to be simulated, or these data may be set at random by the virtual lab. Then, the x(t) curve - displacement versus time- is displayed on a computer screen. Since identifying the parameters of the oscillation is based on its maximum amplitudes, to obtain theoretical parameters, the virtual lab identifies the maximums of x(t) and, from these, it computes the logarithmic decrement (increment) to get the damping -whether positive or negative- the system is undergoing, next from this damping the theoretical parameters of the oscillation are automatically deduced. When the operator of the virtual lab activates the "manual mode," he mouse-clicks the extremes of the x(t) curve appearing on the screen. Once 10 successive extremes have been clicked, the module computes the experimental (manual) oscillation parameters. At the end of the experiment, the virtual lab shows the theoretical and experimental results. Suppose the operator has carefully mouse-clicked

---

* Corresponding Author's Email: jmontenegroj@unmsm.edu.pe

In: Complexity and Fractality in Human Change and Beyond
Editor: Nicoletta Sala
ISBN: 979-8-89530-575-1

the extremes of x(t). In that case, his experimental results will be verified as close or equal to the input data and those calculated by the simulator.

Keywords: computers in education, logarithmic decrement, logarithmic increment, damped oscillations, computational physics, virtual experiments, engineering applications

1. Introduction

Identification of the parameters governing an oscillation is critical not only in engineering applications but also in physical sciences. In general, machinery and mechanisms vibrate while they are operating, and the resulting vibration of a set of pieces partaking in a mechanism has not necessarily a known frequency; it is known that buildings and bridges vibrate due to the passing of vehicles due to the wind and due to earthquakes as well. The subject of this chapter is vital in the last-mentioned cases, for in those events, once the oscillations start, they increase their intensity and, then decrease it, so that it is seen that in those occurrences, there is increment and decrement of oscillations. Note that most vibrations occurring in real life are damped because resistance is always opposing them.

When the logarithmic decrement (LD) is applied to a damped oscillation, the reduction in the size of the maximum displacements is used to assess the damping the system is undergoing. This author has previously dealt with applying the logarithmic decrement to find the parameters of damped oscillations (Montenegro, 2016-a; Montenegro 2017-a), and now, in this work, he is reporting the generalization of the technique to identify also the parameters of stimulated (negative damping) oscillations.

The logarithmic decrement allows experimentally finding the damping an oscillating system is undergoing. Once this damping is figured out, it may be used to compute the viscosity acting on the system and, from this, the maximum amplitude of the oscillation, its initial phase, and its frequency. All these magnitudes are referred to as “control parameters” in this work (Montenegro, 2016-a; Montenegro, 2017-a; Balachandran & Magrab; Thompson, 2003).

This work enhances the educational virtual lab previously released by this researcher (Montenegro, 2016-a; Montenegro, 2017-a). The procedure described in this work illustrates how to get a pair of new parameters by interacting by way of mouse clicks on the x(t) curve generated via computer simulation and displayed on the computer monitor; these two new parameters are the initial phase of the vibrations and its maximum amplitude. The extraction of these two additional magnitudes adds evidence of the power of the LD technique to uncover some extra parameters influencing oscillations; it also reinforces the reasons for fostering the usage of computer simulation and its confidence.

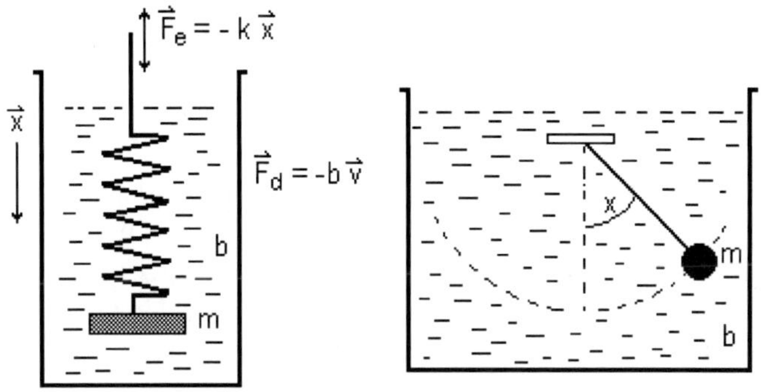

Figure 1. The two main models of damped oscillators, the spring and the pendulum, are both immersed in a viscous medium (viscosity b), which attenuates their vibrations.

In a prior publication (Montenegro, 2016-a), this author reported the creation of a virtual lab which, after displaying on a computer screen the x(t) curve corresponding to the user's input data, makes an automatic analysis of the curve to detect the damping the oscillator is experiencing. From this, the frequency of the vibrations is assessed. Next, in Montenegro (2017-a), an improved and more advanced version of its predecessor was introduced; that new version became highly interactive and friendly because the user was allowed to click with the mouse on the peaks and valleys of the x(t) curve appearing on a computer screen and, the module processed the information from these clicks and, calculated the damping of the oscillation and, from this, the frequency the oscillator was experiencing. This new version (Montenegro, 2017-a) of the Virtual Lab fosters learning through the intense interaction of its user with the computer simulator. Given any two successive extremes of x(t), the LD is mathematically expressed as the natural logarithm of its ratio, and from the fact that these two maximum amplitudes, x_{n+1}, and, x_n are separated by a specific time t, it is supposed that the mentioned ratio satisfies:

$$\frac{x_{n+1}}{x_n} = e^{-\lambda t}, t = t_{n+1} - t_n \quad (1)$$

where λ is a constant to be experimentally determined.

In the last equation, the exponent is negative because, in the event of damped oscillations, the maximum amplitudes continuously decrease. Conversely, this exponent is positive in the case of stimulated oscillations, for in these circumstances, the damping is negative, and maximum amplitudes continuously increase; finally, when the oscillations are uniform -they neither increase nor decrease- the exponent is zero. Equation (1) above applies as long as oscillations are identical; this is uniform, which means that the orbits in State Space remain invariable as time goes forward. In the case of chaotic oscillators, the displacements may describe similar but not equal orbits in state space, the reason why, in the case of chaotic oscillations, the state space looks literally chaotic in the most common sense of the word (Montenegro, 2014; Montenegro, 2016-b). Notwithstanding, the algorithm in this chapter necessarily applies to non-chaotic vibrations; to automatically identify the vibration extremes needed to work out the logarithmic decrement, a technique from chaos theory is applied, namely the Poincaré maps.

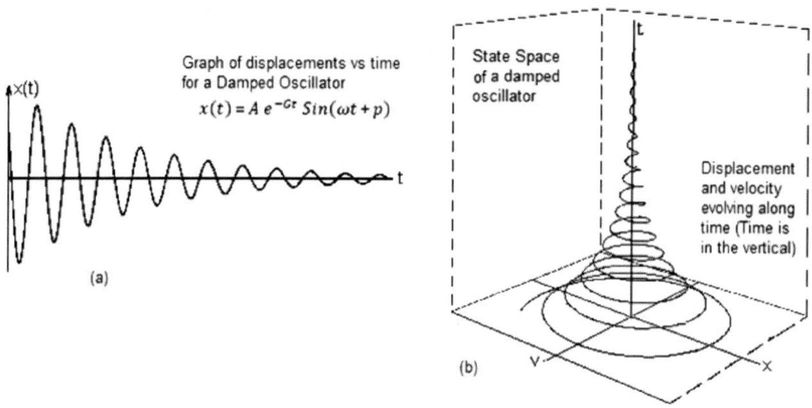

Figure 2. Oscillations of an under-damped (sub-critically damped) vibrator. (a) Its time for evolution. (b) Its 3D state space shows the evolution of displacement and velocity as time progresses.

Figure 2 displays the vibrations of a sub-critically damped oscillator. It is evident in (a) and (b) that due to the viscosity of the environment, the oscillation amplitudes decrease as time goes by until, finally, the vibrations stop. When the oscillations are uniform, as in the cases dealt with in this work, the distance between turns in state space remains constant. If the vibrations are chaotic, they are anything but uniform but behave unpredictably and anarchically.

2. Damped Oscillations

During damped oscillations (see Figures 1 and 2), amplitudes of oscillation reduce gradually as time elapses, and as the damping increases, the contractions of the oscillations become quicker. Considering the connection between the natural frequency of the vibrator and the damping it is experimenting with, damped vibrations may be considered Critical, Subcritical, and Supercritical.

This work deals with subcritical damping, a situation that occurs during vibrations known as under-damped. Figure 1 shows the physicists' two favorite vibrating system models: the vibrating spring and the oscillating pendulum. In both cases, the vibrators are immersed in a medium offering some muffling to the motion; in the simplest case, this medium may be a liquid or a highly dense atmosphere.

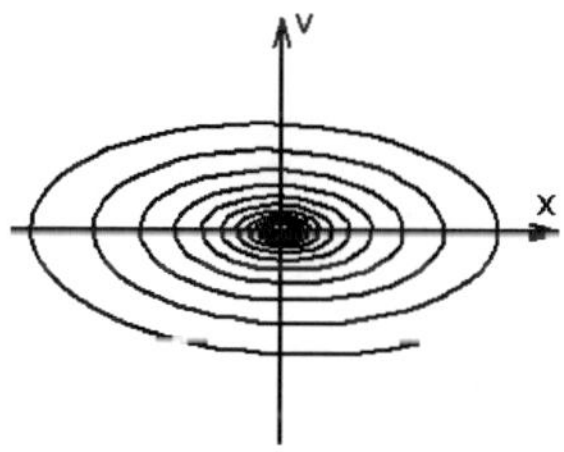

Figure 3. 2D version of the 3D State Space for the case of a damped vibrator.

Figure 3 shows the 2D version of the tridimensional state space for the case of a damped vibrator. The 3D (x, v, t) points have been projected onto the XV plane. The peaks and valleys of x(t) are, respectively, at angles 0 and π with the x-axis, and, in this research, they are used to compute the LD.

3. Differential Equation of Motion under Subcritical Damping

The differential equation of motion of an object undergoing subcritical damping, this is, an under-damped oscillator, is deduced in (Alonso & Finn, 1967; Smith, 2010; French, 1964) and it is given by

$$m\frac{d^2\vec{x}}{dt^2} = -k\vec{x} - b\vec{v} \qquad (2)$$

This equation is studied in engineering elementary physics courses, and it shows the connection between the force acting on the vibrator with the spring's reacting force (Hooke's law) and the viscous damping b exerted by the medium. In the equation above, the minus signs show that the faster the spring oscillates, the higher the resistance due to the damping offered by the environment. Taking into consideration that the velocity is the first temporal derivative of the displacement, the scalar version of the previous equation is

$$\frac{d^2x}{dt^2} + \frac{b}{m}\frac{dx}{dt} + \frac{k}{m}x = 0$$

In this equation, two coefficients are quickly identified:

$$\omega_o^2 = k/m \quad G = b/2m \qquad (3)$$

In equation (3), ω_o is the vibrator's natural frequency, which is the free oscillator's frequency, and G is the muffling (damping), which depends on the viscosity of the environment.

Once ω_o and G are substituted in the differential equation; it may be re-written as

$$\frac{d^2x}{dt^2} + 2G\frac{dx}{dt} + \omega_o^2 x = 0 \qquad (4)$$

In the subcritical damping or underdamped case: $\omega_o > G$.

Since a damped oscillation tends to vanish over time, it can be assumed that the solution of equation (4) has the general form.

$$x(t) = A\,e^{-Gt}\,Sin(\,\omega t + p) \qquad (5)$$

Notice that in eq. (5) it is assumed that the system oscillates with a specific frequency ω, which, as will be seen ahead, is the frequency of the damped oscillations.

Equation (5) implies that the velocity of the oscillations –the first temporal derivative of x(t)- is

$$v(t) = \omega\,Ae^{-Gt}\,Cos(\omega t + p) - G\,x(t) \qquad (6)$$

and that the acceleration –the second temporal derivative of x(t)- is

$$a(t) = \frac{d^2x}{dt^2} = -\omega^2 x - 2G\frac{dx}{dt} - G^2 x \tag{7}$$

Finally, replacing in the differential equation (4):

$$-\omega^2 x - 2G\frac{dx}{dt} - G^2 x + 2G\frac{dx}{dt} + \omega_o^2 x = 0$$

In this way

$$\omega^2 = \omega_o^2 - G^2 \tag{8}$$

This result implies that (5) is a solution of (4) as long as condition (8) is satisfied.

Equation (8) supplies the connection between the natural frequency ω_o , and the frequency ω of the damped oscillations and the damping G experienced by the oscillator. Under these conditions, the period T of the damped oscillations is $T = 2\pi/\omega$. The displacement x(t) and velocity v(t), from (5) and (6) are used to construct the state space of the oscillations.

4. The Differential Equation of the Stimulated Oscillator

In the case of a stimulated oscillator (negative damping), the differential equation is

$$\frac{d^2x}{dt^2} - 2G\frac{dx}{dt} + \omega_o^2 x = 0$$

And making a parallel with equations (4) and (5); in this case, the solution must have the form

$$x(t) = A\,e^{Gt}\,Sin(\,\omega t + p)$$

Then, after replacing the latter expression with the former equation, the following result is obtained:

$$\omega^2 = \omega_o^2 - G^2$$

A relationship has already been obtained (see eq. (8)) for the case of the damped oscillator. In conclusion, equation (8) is valid in both types of oscillation, damped and stimulated.

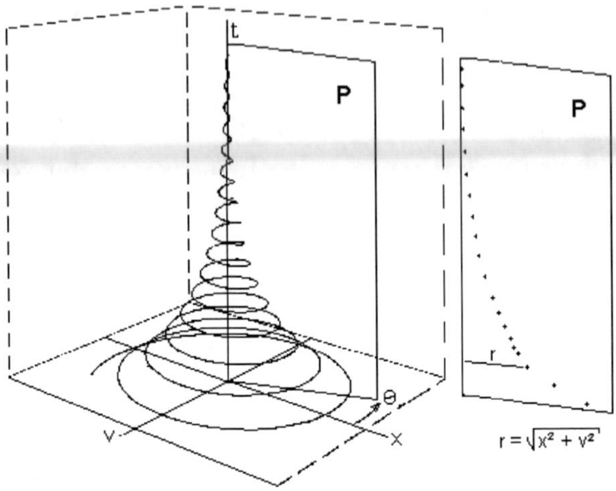

Figure 4. The Poincaré Plane P may be regarded as a longitudinal cut, a cut along the time of the 3D state space.

Figure 4 shows the state space of a damped oscillation; the plane P is the Poincaré map at the angle θ with the x-axis. Notice that this plane may be extracted at any angle between 0 and 2π. Observe that plane P is like a longitudinal tomography of state space. For a given angle with the x-axis, all the intersections of the curve with plane P constitute a Poincaré Map at that angle. Peaks and valleys of x(t), which are used to compute the logarithmic decrement, are at angles 0 and π respectively.

5. The Logarithmic Decrement and the Logarithmic Increment

5.1. The Logarithmic Decrement

In the Logarithmic Decrement, LD (Thomson, 2003; Balachandran & Magrab; Smith, 2010), the relationship (see Eq. 1) between two consecutive maximum amplitudes x_n and x_{n+1} during a damped oscillation leads to

$$x_{n+1} = e^{-\lambda t} x_n \,, \lambda\text{: } Constant \tag{9}$$

In this expression, λ is the logarithmic decrement and points x_n and x_{n+1} are two extremes (either peaks or valleys) in sequence in the curve of displacement versus time (see Figure 2), and they may be obtained by extracting Poincaré Maps of phase space, as shown ahead. Notice that the peaks of x(t) are the points at 0° with the x-axis (see Figure 3), and valleys are those at 180°.

It is very well known from oscillation basic theory that the period T of the oscillation is the time elapsed between any two consecutive points with the same phase in the x(t) curve. Hence, the previous equation may be rewritten as

$$\frac{x_{n+1}}{x_n} = e^{-\lambda T}$$

and from this expression:

$$\lambda = -\frac{1}{T}\ln\left[\frac{x_{n+1}}{x_n}\right] \tag{10}$$

where T is the period of the muffled oscillations, and it is given by $T = 2\pi/\omega$. At experiment time, the value of λ in the equation above, it is assessed by averaging several consecutive cases of equation (10). The application of equation (5) to two consecutive oscillation extremes separated by a period T yields

$$\begin{cases} x_n = A\,e^{-Gt}\,Sin(\omega t + p)\ (11) \\ x_{n+1} = A\,e^{-G(t+T)}\,Sin(\omega(t + T) + p) \end{cases} \tag{12}$$

In this way, from equation (12)

$$x_{n+1} = A\,e^{-G(t+T)}\,Sin(\omega t + p + \omega T)$$

and since $\omega = 2\pi f = 2\pi/T$, then:

$$x_{n+1} = A\,e^{-G(t+T)}\,Sin(\omega t + p + 2\pi)$$
$$x_{n+1} = A\,e^{-G(t+T)}\,Sin(\omega t + p) \tag{13}$$

So that from equations (11) and (13):

$$\frac{x_{n+1}}{x_n} = e^{-GT}$$

and extracting G:

$$G = -\frac{1}{T} \ln\left[\frac{x_{n+1}}{x_n}\right] \quad (14)$$

By operating on equations (14), (10), and (8):

$$\lambda = G = \sqrt{\omega_o^2 - \omega^2} \quad (15)$$

Be aware of the fact that from the latest equation and equation (3), the viscosity $b = 2mG$ of the environment may be assessed.

Equations (10) and (14) are similar to the Lyapunov exponent equation from Chaos Theory, where the value of the maximum exponent signals the degree of chaos in a chaotic system (Montenegro, 2016-c; Montenegro, 2017-b). In chaos research, in order to assess the Lyapunov exponents, two initially neighboring state space orbits are analyzed, whereas, when assessing the logarithmic decrement, a single state space curve is investigated.

5.2. The Logarithmic Increment

In the case of a stimulated vibrator (negative damping), the amplitudes are every time larger, so it can be assumed that the displacement is given by

$$x(t) = A\, e^{Gt}\, Sin(\omega t + p)$$

and it can also be assumed that two consecutive extremes of these oscillations satisfy

$$x_{n+1} = e^{\lambda T}\, x_n$$

Then, by a procedure similar to that carried out to obtain equation (14), it can be shown that:

$$\lambda = G = \frac{1}{T}\ln\left[\frac{x_{n+1}}{x_n}\right]$$

6. The Initial Phase and Maximum Amplitude of the Oscillations for Damped and Stimulated Oscillations

6.1. Damped Oscillations

From equations (5) and (6), the initial values of displacement and velocity are

$$t = 0 : \begin{cases} x_o = A\,Sin\,p \\ v_o = \omega A\,Cos\,p - Gx_o \end{cases}$$

In this way

$$\begin{cases} Sin\,p = x_o/A \\ Cos\,p = (v_o + Gx_o)/\omega A \end{cases}$$

Thus, the initial phase of the oscillations is

$$Tan\,p = \frac{\omega\,x_o}{v_o + G\,x_o} \Rightarrow p = Tan^{-1}\left[\frac{\omega\,x_o}{v_o + G\,x_o}\right] \tag{16}$$

Moreover, the initial displacement (Amplitude) is:

$$A = \sqrt{x_o^2 + \left(\frac{v_o + Gx_o}{\omega}\right)^2} \tag{17}$$

6.2. Stimulated Oscillations

In this case, the equations of motion and velocity are

$$x(t) = A\,e^{Gt}\,Sin(\,\omega t + p)$$

$$v(t) = \omega\,Ae^{Gt}\,Cos(\omega t + p) + G\,x(t)$$

where the amplitude of the oscillation and the initial phase are now given by

$$A = \sqrt{x_o^2 + \left(\frac{v_o - Gx_o}{\omega}\right)^2} \; and \; p = Tan^{-1}\left[\frac{\omega\, x_o}{v_o - G\, x_o}\right]$$

7. Poincaré Maps: A Technique from Chaos Theory

Though this section is not important from the users of the virtual lab point of view, it is included because it is fundamental at software development time since the computer uses this technique to detect the extreme amplitudes (peaks and valleys) of the oscillations, which are used to calculate the logarithmic rate between every pair of consecutive maximum displacements. In studies of chaos based on computer simulation (Montenegro, 2016-c; Montenegro, 2016-d), the detection of peaks and valleys, this is, the extremes of the x(t) curve -which are used in the calculation of the logarithmic decrement- may be achieved by extracting the Poincaré maps –a technique from chaos theory- at 0° and 180°, in that order. It is due to this reason that for someone who has developed computer programs to extract Poincaré Maps in chaotic systems, the logarithmic decrement is not so extraordinary.

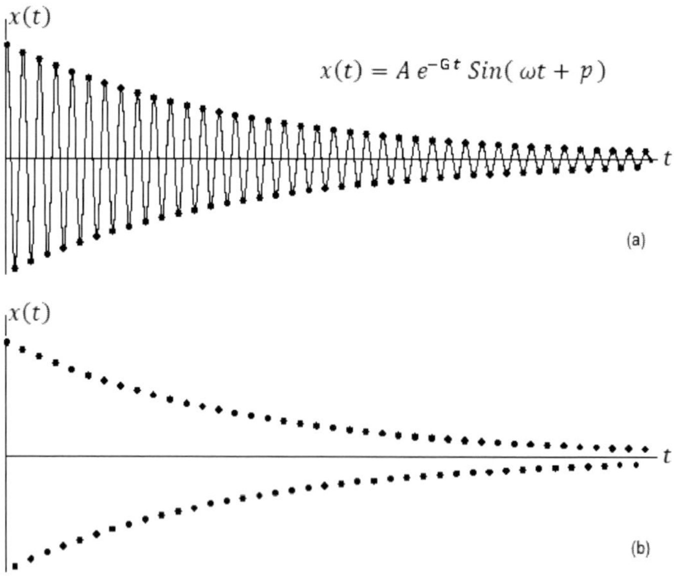

Figure 5. (a) The curve of x(t) for a damped vibrator that appears on the computer screen (b) The corresponding highlighted peaks and valleys.

The here-reported simulation software operates in two stages, namely "Automatic" and "Manual." The Poincaré maps technique is used only while in the Automatic Stage, to identify the peaks and valleys (extremes) of the x(t) curve, which is generated as part of the operation of the virtual lab; this means that the operator of the simulator does not require to be aware of this technique, because it is applied by the software.

In practical applications, when it is possible to identify the extremes of the x(t) curve by any other method, there is no need to extract any Poincaré map. This will be appreciated in the section dealing with the "Manual Stage," where the x(t) curve is displayed on a computer monitor, and its extremes are -at simple sight- identified by the virtual lab operator and are mouse-clicked.

Figure 5 displays the curve of x(t) at the top for a damped oscillator. The simulation module being reported renders this curve on a computer screen. At the bottom, peaks and valleys (dots on curve extremes) have been highlighted during the automatic stage of the execution and after extracting the Poincaré Maps at 0° and 180°, respectively.

7.1. Some Previous Applications of the Logarithmic Decrement

Magalas (Magalas, 2006; Magalas & Majewski, 2008) introduces a new algorithm to assess the logarithmic decrement with high accuracy, and his algorithm applies to cases of rather high damping. In the case of vibrations produced by earthquakes in civil engineering structures, like buildings, Butterworth et al., (Butterworth et al., 2004) report finding the application of the Logarithmic decrement useful. In the mechanical spectroscopy of solids, Majewski and Magalas apply the logarithmic decrement (Majewski & Magalas, 2015).

7.2. Performing the Computer Simulation Experiment

Figure 6 displays a schematic representation of the whole process achieved by the virtual lab being reported. The flowchart shown is an improvement of the one previously published by this author (Montenegro, 2017-a), and it allows the interested reader to visualize – and, if desired, reproduce- the procedure and computations.

This work's simulation software operates under two modes: automatic and manual (or experimental). While in automatic mode, the software automatically detects the peaks and valleys of x(t) and computes theoretical results. On the other hand, when the software runs under manual mode, the user directly interacts with the computer, generating in this way experimental results. Once a simulation finishes, the user may compare theoretical and experimental results.

Right before starting an experiment, the user enters m, k, and b, which are, respectively, the mass of the vibrator, its elastic constant, and the viscosity of the environment. In addition, the simulator may receive, as input data, the maximum amplitude of the vibrations, their initial phase, and the time interval (time-step) to be used in the simulation.

Setting parameters like time-step, maximum amplitude of oscillation, and initial phase is not essential for the operation of the simulator because they have pre-defined default values in the software; however, since this is an educational software, the user may fix them to see their effect in the simulation. At this point, it is important to remember that the larger the maximum amplitude, the more evident the peaks and valleys of x(t).

Though the flowchart in Figure 6 deals with the logarithmic decrement (damped oscillations), it can be easily adapted for stimulated oscillations (negative damping).

7.3. Automatic Stage

About the flowchart in Figure 6, right after input parameters are provided to the simulation module by the user, theoretical parameters like the free oscillator angular frequency ω_0 and the damping G of the medium are automatically calculated. To accomplish this, the virtual lab applies Equation (3). In addition, by application of Equation (8), the angular frequency of the damped oscillations ω is calculated together with the period T of the oscillations. Next, with the help of equations (5) and (6), both the initial displacement x_o and initial velocity v_o are obtained.

At this point, Equation (5) is run to generate several oscillation periods, ensuring enough peaks and valleys are shown on the computer screen so that Equation (10) may be assessed. To average an acceptable number of peaks and valleys, a number of 10 peaks and 10 valleys must be separately averaged to manually calculate λ.

Once the computer simulation starts, the curve of displacement versus time x(t) and the state space are both shown on the computer monitor (Figures

3 and 5). Next, the system automatically extracts Poincaré maps at angles 0 and π in state space, this operation highlights the peaks and valleys of x(t) (Montenegro, 2014; Montenegro, 2016-b; Montenegro, 2016-c; Montenegro, 2017-b). At this point, the value of λ in equation (10) is assessed from every pair of successive peaks (valleys), and with the help of equation (15), the values of G and ω are both obtained. During the manual stage of the simulation, some 10 peaks and -optionally- 10 valleys of x(t) are required to manually calculate λ.

7.4. Manual Stage

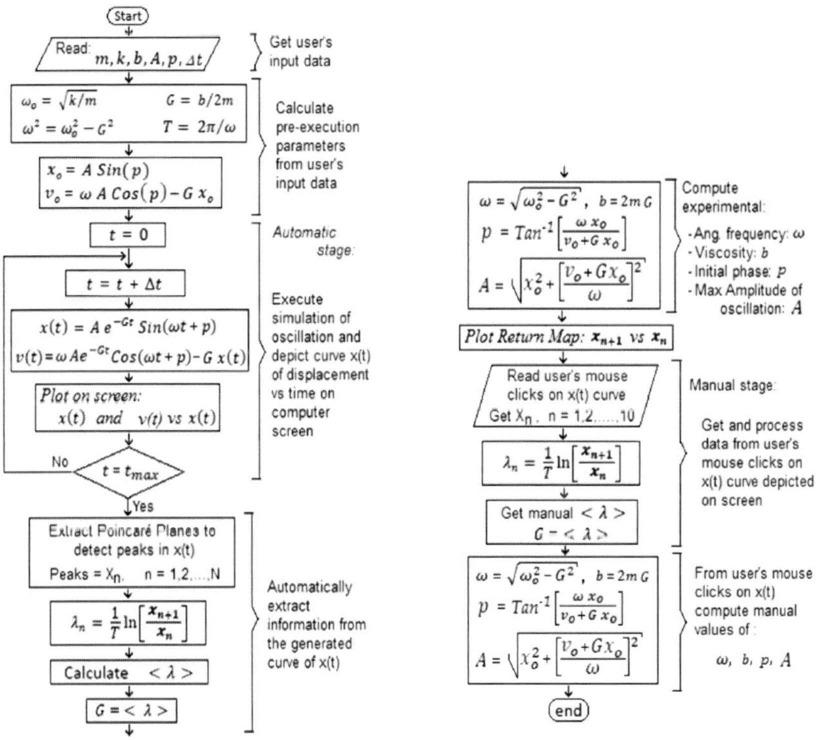

Figure 6. A flowchart of the LogDec-LogInc computer program is described.

While the simulator is under manual mode (see Figure 6), the user may identify at simple sight 10 successive peaks of the x(t) curve appearing on screen, and once he mouse-clicks each of them, the simulator assesses their average logarithmic decrement λ and the damping G the oscillator is undergoing; additionally, other oscillation parameters are computed at this time.

These lately computed data are close – when not equal - to the corresponding theoretical values computed at the outset of the simulation.

Optionally, the user may repeat the operation described above, this time by mouse-clicking the valleys of the displacement versus time curve, x(t), and the resulting data must be -as previously described- close to the theoretical parameters assessed right at the outset of the experiment. There is no particular order; the user may first click the valleys of x(t) and then its peaks or start the manual stage by clicking the peaks and then the valleys.

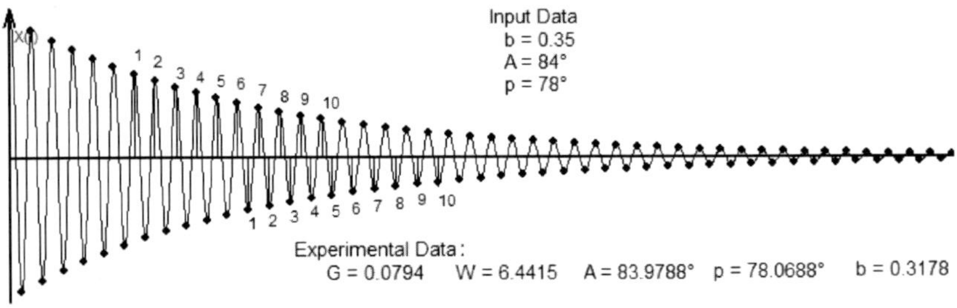

<><> Virtual Lab: Log Dec - Log Inc <><>
Experiment N = 1
INPUT DATA .....
Decreasing Osc. - Log Decrement
Mass m = 2 kg
Elastic constant k = 83 N/m
Viscosity b = 0.3500
Amplitude A = 84°
Initial phase p = 78° = 1.3614 rad
Delta time = 0.0250 s

AUTOMATIC STAGE ......

Theoretical parameters:
Damping of the oscillator G = 0.0875
Natural Angular frequency Wo = 6.4420 rad/s
Damped angular frequency W = 6.4415 rad/s
Damped Oscillations period T = 0.9754
Initial displacement Xo = 82.1644
Initial Velocity Vo = 105.3079
N time-steps = 2500
Elapsed time during oscillation = 59.900

Automatic calculation of Logarithmic Decrement
Damping: G = 0.0875
Damped Angular frequency: W = 6.4415
Amplitude A = 83.9999°
Initial Phase p = 78.0002° = 1.3614 rad
Viscosity b = 2 m G = 0.35

MANUAL STAGE .....

Results after mouse clicking the peaks of X(t) curve ....
Damping: G = 0.0937
Damped Angular frequency: W = 6.4414
Amplitude A = 84.0165°
Initial phase p = 77.9473°
Viscosity b = 2 m G = 0.3747

Results after mouse clicking the valleys of X(t) curve ...
Damping: G = 0.0652
Damped Angular frequency: W = 6.4417
Amplitude A = 83.9412°
Initial phase p = 78.1903°
Viscosity b = 2 m G = 0.2609

Average results from clicking peaks and valleys .....
Damping: G = 0.0794
Damped Ang. frequency: W = 6.4415
Amplitude: A = 83.9788°
Initial phase: p = 78.0688°
Viscosity: b = 0.3178

Figure 7. The output of a virtual experiment with damped oscillations.

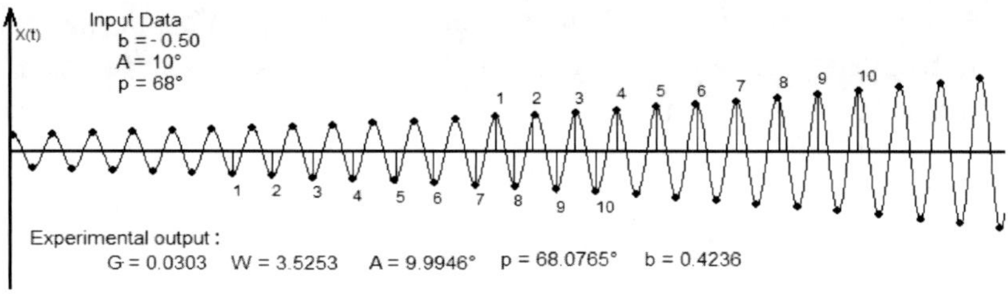

<><> Virtual Lab: Log Dec - Log Inc <><>
Experiment N = 1
INPUT DATA.....
Stimulated Osc. - Log Increment
Mass m = 7 kg
Elastic constant k = 87 N/m
Viscosity b = 0.5000
Amplitude A = 10°
Initial phase p = 68° = 1.1868 rad
Delta time = 0.0250 s

AUTOMATIC STAGE ......
Theoretical parameters:
Damping of the oscillator G = 0.0357
Natural Angular frequency Wo = 3.5254 rad/s
Damped angular frequency W = 3.5252 rad/s
Damped Oscillations period T = 1.7823
Initial displacement Xo = 9.2718
Initial Velocity Vo = 12.8746
N time-steps = 2500
Elapsed time during oscillation = 61.600

Automatic calculation of Logarithmic Decrement
Damping: G = -0.0357
Damped Angular frequency: W = 3.5252
Amplitude A = 9.9312°
Initial Phase p = 69.0049° = 1.2044 rad
Viscosity b = - 0.50

MANUAL STAGE .....

Results after mouse clicking the peaks of X(t) curve ....
Damping: G = 0.0293
Damped Angular frequency: W = 3.5253
Amplitude A = 9.9937°
Initial phase p = 68.0900°
Viscosity b = - 0.4102

Results after mouse clicking the valleys of X(t) curve ....
Damping: G = 0.0312
Damped Angular frequency: W = 3.5253
Amplitude A = 9.9956°
Initial phase p = 68.0631°
Viscosity b = - 0.4371

Average results from clicking peaks and valleys .....
Damping: G = 0.0303
Damped Ang. frequency: W = 3.5253
Amplitude: A = 9.9946°
Initial phase: p = 68.0765°
Viscosity: b = - 0.4236

Figure 8. The output of a virtual experiment with negative damping (stimulated oscillations).

In any case, results from the two measurements may be averaged to improve statistics. The only restriction at this stage is that peaks or valleys clicked must be successive.

Figures 7 and 8 display the outputs generated by the reported virtual lab during experiments with damped and stimulated oscillations, respectively. At the top of these figures appear the screenshots of x(t) displayed on the computer screen; the small numbers from 1 through 10 at the peaks and valleys of x(t) are the user's mouse clicks. At the bottom, the experimental reports generated by the simulator are shown.

A verification of the found control parameters -which may be achieved at the end of the simulation- confirms that they are very close –when not equal- to the corresponding theoretical values. In theory, it is not required to click on exactly 10 peaks or valleys of the displacement versus time curve; clicking on just two or three is sufficient.

However, with the aim of higher accuracy when computing the average, the software has been made so that the user must mouse-click 10 peaks and -if desired- 10 valleys of x(t).

8. Demonstrative Execution of the Virtual Lab

8.1. Damped Oscillations

Figure 7 exposes an example associated with damped oscillations; the upper part of this figure displays the x(t) curve associated with the simulation report displayed on the bottom of it, whose input data are: b = 0.35, A = 84° and p = 78°. After averaging the results from the user's mouse clicks on peaks and valleys, the output is:

$$G = 0.0794, \omega = 6.4415, A = 83.9788^{\circ}, p = 78.0688^{\circ}, b = 0.3178$$

Concerning the lower part of Figure 7, at the start, when the user enters the input data, the software computes the so-called "theoretical parameters," this is followed by the automatic detection of peaks and valleys of the oscillation.

From these, the logarithmic decrements, as well as some control parameters, are obtained. Next, during the manual stage, when the user mouse-clicks the peaks and valleys of x(t) appearing on the screen, some "experimental" control parameters are obtained. This procedure allows the user to compare theoretical results with experimental ones easily.

8.2. Oscillations without Damping

When executing a simulation without damping, the oscillations –as expected-are uniform; that is, they neither contract nor expand, and the simulator detects the expected control parameters.

This means that the technique of logarithmic decrement also operates successfully when there is no damping at all.

8.3. Oscillations with Increasing Amplitudes, Negative Damping

Figure 8 exposes an example with negative damping (stimulated oscillations); the top of this image shows the oscillations associated with input data: b = -0.50, A = 10° and p = 68°. Next, after averaging the results from the user's mouse clicks on peaks and valleys, the following output is obtained:

$$G = 0.0303, \omega = 3.5253, A = 9.9946^{\circ}, p = 68.0765^{\circ} \; and \; b = -0.4236.$$

8.4. Some Experimental Results

Figure 9 exhibits some experiments whose control parameters were randomly set. The graphs show some input data (viscosity, amplitude, and initial phase) to run the simulation, and the resulting displacement curve vs time x(t) is displayed on a computer screen.

The numbered points from 1 through 10 are the user's mouse clicks on the peaks and valleys of x(t). Each graph displays the output parameters resulting from these mouse clicks. The more accurate the position of the mouse clicks at the extremes of x(t), the better the result will be.

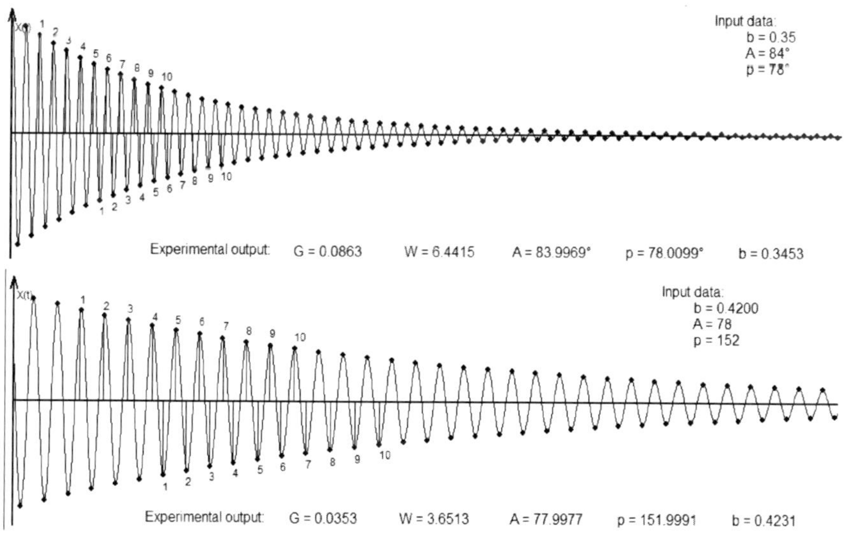

Figure 9. (Continued).

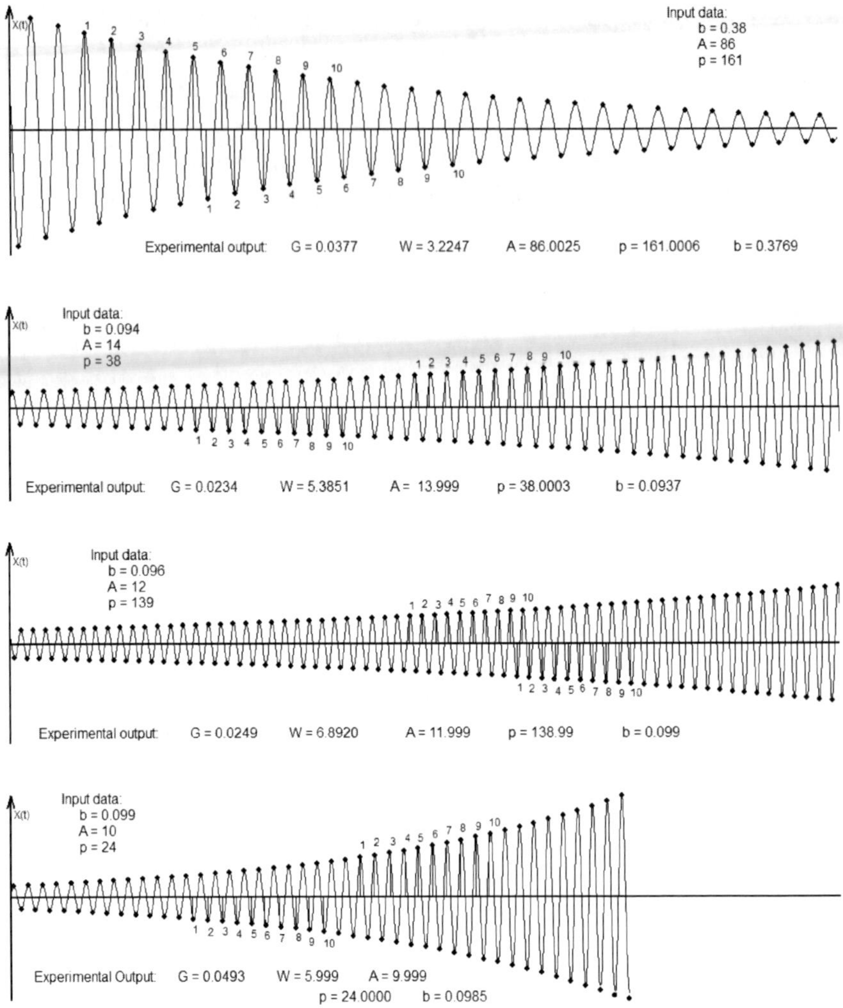

Figure 9. Some results from test executions of the virtual lab.

Conclusion

This chapter outlines the development of an interactive educational virtual lab designed to identify the parameters of oscillation, whether damped or stimulated. The theoretical foundation supporting the software is the Logarithmic Decrement technique.

The most important lesson from this virtual lab is that in everyday life, to identify the parameters of an oscillation, some of its peaks (or valleys) can be measured, and then, by applying the aforementioned technique, its control parameters can be determined. It has been shown that some computational tools from chaos theory, like the extraction of Poincaré maps, may be used only during computer simulation to automatically detect the peaks and valleys of the oscillation amplitudes, which are used to feed the Logarithmic decrement (increment) equations. When computer simulation is not used to assess the logarithmic decrement (increment), the Poincare maps are not required. It has also been proven that the reported virtual lab computes the factual oscillation parameters even when there is no viscosity and damping is null.

The reported virtual lab makes an extension of the logarithmic decrement technique, usually applied to identify the control parameters of damped oscillations, to the case of vibrations under negative damping, which implies increasing vibration amplitudes, and it has been encountered that the technique operates successfully in both cases, including the case of no damping at all.

This work has provided sufficient information and details for interested readers to develop their computer simulators.

References

Alonso, M., & Finn, E. J. (1967). *Fundamental University Physics*, Boston: Addison-Wesley.

Balachandran B., & Magrab E., (2004). *Vibration,* Pacific Grove, CA: Brooks-Cole Editors, 145-148.

Butterworth, J., Lee, J. H., & Davidson B. (2004). Experimental determination of modal damping from full scale testing, *13th World Conference on Earthquake Eng.*, Vancouver, B. C., Canada, Paper No. 310.

French, A. (1964). *Vibrations and Waves*, New York: W. W. Norton & Co.

Magalas, L. B. (2006). Determination of the Logarithmic Decrement in Mechanical Spectroscopy. *Solid State Phenomena,* Vol 115.

Magalas, L. & Majewski M. (2008). Recent Advances in Determination of the Logarithmic Decrement and the Resonant Frequency in Low-Freq. Mech. Spectroscopy. *Solid State Phenomena* Vol. 137, 15–20.

Majewski, M. & Magalas L. (2015). Critical assessment of the issues in the application of Hilbert transform to compute the logarithmic decrement, *Archives of Metallurgy and Materials* 60(2), 1105–1108.

Montenegro, J. J. (2014). Multiplicity and Transitoriness of Chaotic Events, *RIF-UNMSM* 17(2), 1-6.

Montenegro, J. J. (2016-a). Use of the Logarithmic Decrement to Assess the Damping in Oscillations. *Revista de Investigación de Física* (RIF), Vol. 19(1), 1-5.

Montenegro, J. J., (2016-b). Variety and Finiteness of Chaotic Events in the NLDF Oscillator, *Chaos and Complexity Letters*, 10 (1), 81-87.

Montenegro, J. J. (2016-c). The Higher the Damping, the Lower the Chaos Intensity, *Chaos and Complexity Letters,* 10(3), 253-260.

Montenegro, J. J. (2016-d). Velocity behavior during chaotic events in the NLDF Oscillator, *Chaos and Complexity Letters*, 10 (2), 173-185.

Montenegro, J. J. (2016-e). Diversity and temporality of chaotic events, *Industrial Data* 19(1), 125-131.

Montenegro, J. J. (2017-a). Virtual Lab to run Logarithmic Damping Decrement Experiments. *Internat. Journal of Science and Research* (IJSR), Vol 6(2), 1221-1226.

Montenegro, J. J. (2017-b). Viscosity levels and strength of chaos, *International Journal of Science and Research* (IJSR), 6(1). 1020-1024.

Smith, W. (2010). *Waves and Oscillations*, Oxford: Oxford University Press.

Thomson, W. T. (2003). *Theory of vibration with applications*, Burgh le Marsh, Lincolnshire, United Kingdom: Stanley Thornes Publisher.

About the Editor and Authors

Editor

Nicoletta Sala received a degree in Physics ("Laurea") and applied cybernetics at the University of Milan (Italy). PhD in Communication and Information Science at the University of Lugano (USI, Lugano, Switzerland). Postgraduates (two years for each) in: "Didactics of the Communication and Multimedia Technologies" and "Journalism and Mass Media." She is an Information Technology and Electronics professor, and she taught at the University of Lugano (Mendrisio, Switzerland), at the University of Insubria (Varese, Italy), and at the European School of Economics (London, UK). She was the founder and editor of *Chaos and Complexity Letters International Journal of Dynamical System Research* (Nova Science Publisher, New York). She is one of the founders and editors of the new book series *Human Dynamics: Multidisciplinary Complexity Science* (Nova Science Publisher, New York). Her research interests concern various scientific topics from an interdisciplinary point of view and comprise the following areas: Fractal Geometry and Complexity, Mathematics in Arts, Architecture, and Industrial Design; New media and IT in the learning environments; and Virtual Reality in education. She has authored 16 mathematics and information technology books and edited 19 others. She has written 370 scientific works. She is the founder and professor of Istituto di Ricerca in Fisica Aritmetica (Research Institute in Arithmetic Physics, IRFIA, Switzerland).

Authors

Massimo Agnoletti is a psychologist with a PhD and an expert in Stress, Positive Psychology, and Epigenetics. He has been involved in clinical and scientific research for over twenty years and is an international lecturer. He has produced more than a hundred scientific publications.

Laura Cancellara is a clinical psychologist, and doctor in motor and sports activities science. She is an Experienced Teacher with a demonstrated history of working in the e-learning industry. Skilled in Coaching, Event Management, English Management, and Marketing. Strong professional education with a "Laurea Specializzazione Universitaria" (degree) focused in Suor Orsola Benincasa of Napoli (Naples, Italy) from "Specializzazione Scienze dell'Educazione" (Master in Educational Sciences). She is the Project Manager of Mind Force Society and a member of the Institute for Complexity Studies (Rome, Italy).

Gabriele Cappellato received his first degree at "Istituto Universitario di Architettura di Venezia (IUAV, Venice, Italy) and a PhD. in architectural design at IUAV. He collaborates with several Italian and international architectural firms, participating in studies, competitions, projects, and creations. He has curated many exhibition installations for important venues in Italy and abroad. His study and research activities are intense, focusing on in-depth studies of modern and contemporary architecture. He has collaborated with magazines and with the leading national and international periodicals. He has been a contract professor at several universities, including Ferrara, Milan, IUAV, and Padua. From 1995 to 2018, he was a design professor, dean of students, and head of bachelor's at the Academy of Architecture of Mendrisio of the University of Italian Switzerland. He is now an Emeritus Professor.

Gabriele Losa gained international renown through pioneer papers on fractal geometry applied to cell biology and human cancers. He was a Member of the European Academy of Sciences. BSc in Natural Sciences and PhD (1972) from the Faculty of Biology and Medicine of Lausanne University. Postdoctoral Fellow at the Department of Cell Biology and Clinical Cancer Research Institute of Berne University, the Institute of Cellular Pathology of Louvain University, and the Memorial Sloan Kettering Cancer Center of New York. Founder of the Laboratory of Cellular Pathology and the Institute for Scientific Interdisciplinary Studies, Locarno (Switzerland). He taught cell biology at the Universities of Lausanne, Turin, Bari, and Naples. Academic collaboration with the Department of Mathematical Physics of the University of Ulm aimed at investigating the role of fractal geometry in biology and medicine and developing mathematical software for fractal analysis.

Danilo Merlini received his first degree in Physics (Master of Science) at ETH Zurich (Switzerland) and a PhD in Sciences (Theoretical Physics) at École Polytechnique Fédérale (EPFL) de Lausanne (Switzerland). He carried out his research at universities in Germany and the USA. He is a professor of Mathematics and Physics. His research areas include Number Theory, Applied Mathematics, Mathematical Analysis, Physics, Thermodynamics, and Phase Transitions. He is the founder, professor, and director of the Centro di Ricerca in Fisica e Matematica (Research Center for Mathematics and Physics, Locarno, Switzerland); founder and professor of the Istituto di Ricerca in Fisica Aritmetica (Research Institute in Arithmetic Physics, IRFIA, Switzerland); and founder and member of the Institute for Scientific Interdisciplinary Studies, Locarno (Switzerland). He is also a co-organizer of international congresses and conferences on Fractal Geometry, Complexity, and Physics.

Javier Montenegro Joo is a physicist specializing in computational and simulation-based research. He is a physics faculty at San Marcos University (Lima-Perú) and founding director of the VirtualDynamics Org., a development institution. He creates Virtual Labs on science and engineering; he has developed virtual labs on Stochastic Aggregation, Cellular Automata, Percolation, Chaos, Fractals, Neural Networks, Genetic Algorithms, and Pattern Recognition. His most renowned brainchild and commercial success is the Physics Virtual Lab (PVL), an ultra-interactive collection of 203 general physics simulation modules used to form science and engineering students. He also devises algorithms to create astonishing algorithmic artworks.

Gabriele Serafini is Associate Professor of History of Economic Thought at Niccolò Cusano University, Rome, Italy. Director and Scientific Head of the Dynamic and Behavioral Economics Area at "CESDE" - Centre for the Analysis of Economic Dynamics, his scientific interests focus on theories of production, exchange, and distribution of wealth, both in relation to their internal coherence and their historical dimension.

Francesca Stecchi is a Clinical and Sport Psychologist. She holds a degree in Clinical and Community Psychology from the University of Padua (Italy), and she has completed three master's programs in Sport Psychology: two in Spain and one in Italy. Founder of Insynergy – Sport Psychology. She is President of the Veneto Regional Committee of the Italian Federation of Sport Psychologists (FIPsiS). Member of the Sport Psychology Working Group of the Veneto Psychologists' Association. Mental preparation consultant for the Italian Tennis and Padel Federation (FITP) in Venice and its province. Head of Olympic and pre-Olympic mental preparation for the Italian National Canoe Sprint Team. Member of the FIPAV Veneto staff as a trainer and sports psychologist. Staff member of Pallavolo Padova Kioene (Italy). She primarily works with professional and elite athletes in various disciplines and collaborates with clubs, Federations, and the "Fiamme Oro" sports group.

Mohamadreza Tamamgar is currently working at the Department of Mathematics, Islamic Azad University of Gonabad (Gonabad, Iran). He is active in applied mathematics, complexity, and dynamical systems research. One of his recent publications is *A New Reconstruction of Variational Iteration Method and Its Application to Nonlinear Volterra Integrodifferential Equations*.

Index

E

F

G

H

I

L

M

N

O